A CENTURY OF RESEARCH

Centennial History of the Arkansas Agricultural Experiment Station

by
Stephen F. Strausberg

Arkansas Agricultural Experiment Station
Fayetteville, Arkansas 72701

On the Cover

A tower of the original Agricultural Experiment Station building, built in 1888 and razed in 1972. Illustration and cover design by Corinnia Briggs. *(See photograph on page 7.)*

Published by
Arkansas Agricultural Experiment Station
University of Arkansas

Gerald J. Musick, Director
Preston E. La Ferney, Vice President for Agriculture

Nancy Griffith Wyatt, editor
Raymond W. Barclay, Jr., photo editor

Library of Congress Catalog Number: 89-6951
ISBN: 0-9622858-0-3
Special Report 136
EP3M290PM

CONTENTS

ACKNOWLEDGMENTS

I wish to thank the many good people of the Arkansas Agricultural Experiment Station who have entrusted their important story to a historian from New York City. Without their support and encouragement, this "city boy" could never have completed this project.

In particular, I want to acknowledge the good humor and assistance of Ray Barclay, without whom I would have never attempted this project; Dr. Don Voth and Dr. Gerald Musick, who conceived the original idea of a Centennial history; Eloise Cole, who showed enormous patience in deciphering my handwriting. I want to show my deep appreciation for the efforts of Dr. L.O. Warren and Dr. C.J. Brown, who reviewed the manuscript and offered their insightful criticism.

The list of scientists at the Station who consented to my sometimes rambling interviews is a long one, but I want to particularly thank Dr. L. Thompson, Dr. C. Caviness, Dr. C. Berry, Dr. J. Moore and Dr. M. Redfern for directing my efforts.

I would also like to thank Nan Lawler and Charlene Kaufman in Special Collections at the University of Arkansas Library for their assistance locating photographs and Gary Shepherd in Media Services for his help in copying them.

Any history is ultimately a collective endeavor. Consequently, Nancy Griffith Wyatt deserves special mention as an insightful and intelligent editor of the manuscript.

Finally, I want to thank my wife Michaela and my two daughters, Brocha and Rachel, who have tolerated the piles of notes, books and papers that have littered my house during this two-year endeavor.

Stephen F. Strausberg
Fayetteville, Arkansas
May 1989

PREFACE

In 1888 Albert Menke was appointed the first director of the newly created Arkansas Agricultural Experiment Station. He had the awesome responsibility and task to select a team of scientists to study the problems facing Arkansas farmers, stock raisers and fruit growers and to make recommendations to improve production. Under Menke's leadership, agricultural research based on science began in Arkansas. The studies conducted by Menke and his staff of 12 researchers, including a botanist, biologist, horticulturist, entomologist, veterinarian, three chemists and four assistants, would lay the foundation that has lasted 100 years. Today, Menke's Agricultural Experiment Station is Arkansas' premier agricultural research institution.

This book tells the story of the men and women who applied their talents to agriculture, home economics and forestry in service to the people of Arkansas. It recounts their accomplishments, their failures and their disappointments. The author, Stephen Strausberg, brings to light the Station's successful tick eradication program during the early 1900s and the struggle to keep the Experiment Station in Fayetteville. He discusses the contributions of researchers such as entomologist Charles Woodward, considered the father of pest management in Arkansas, and M.C. Kick, who developed the process of parboiling rice.

Strausberg tells us about Barnett Sure, who actually discovered vitamin E but is recognized as a co-discover because he didn't publish his findings until several years later. He also reviews the accomplishments of the Station's first veterinarian, R.R. Dinwiddie, who helped develop the first vaccine for swine cholera and then travelled the dirt roads of the Ozarks helping farmers vaccinate animals.

After reading "A Century of Research," I feel proud and humble to serve in the same office as Albert Menke did 100 years ago and to follow in the footsteps of the Station's other 12 directors. The responsibilities of the office have grown and the challenges continue. I hope during my tenure to bring to the office the energy, enthusiasm and foresight of my predecessors who not only charted the beginning of Arkansas agriculture but provided incredible guidance during some very difficult and trying times. They are to be commended for their vigilance and perseverance in the face of incredible adversity.

Our future challenge is to provide the Station's faculty and staff with the resources to explore and to encourage them to take the risks necessary to investigate and discover the new technologies and scientific principles that will be required to insure the continued success of agriculture. We must continue to provide Arkansans and the citizens of this nation with unbiased, accurate and essential information if we are to solve the problems of tomorrow.

I am pleased to write the preface to this book, and I wish to thank Steven Strausberg who did a remarkable job of compressing 100 years of research into words and pictures that tell the story of a century of agricultural research in Arkansas.

Gerald J. Musick, Director
Arkansas Agricultural
Experiment Station

Chapter I

The Early Years of the Experiment Station

The roots of the Arkansas Agricultural Experiment Station and the promotion of the application of science to farming can be found in the early history of the state. However, the use of scientifically tested farming techniques in a state still marked by primeval forests, swampy lowlands and unexplored remote valleys presented many challenges. The pioneer stock emigrating into Arkansas had to carve out a niche, establish farms and maintain a living before they could turn their attention to implementing the new agricultural techniques. Today, Arkansas agriculture is among the most efficient in the world, a leader in soybeans, rice and poultry. The inspiring story of the untiring efforts of scientists to overcome obstacles and develop a viable, productive agricultural sector is a wonderful example of tenacity and ability in the face of financial and geographic problems.

The story of scientific agricultural research in Arkansas began in 1842 when Governor Archibald Yell asked the state legislature to appropriate money for research. Readers of prominent southern journals of the time, such as *DeBow's Review*, knew that even in Arkansas with its vast tracts of virgin land, soil depletion was an ever-looming problem. The older states of the southeast, Virginia and South Carolina, were already marred by gully erosion and soil exhaustion. Nevertheless, the people of pioneer Arkansas, with rich delta and river valley land that had barely felt the farmer's plow, could scarcely conceive of the eventual difficulties and challenges that today's farmers face.

The passage of the Morrill Land-Grant College Act in 1862 by a Republican-controlled Civil War Congress was in response to the decline of agriculture in New England and in the face of competition from the newly opened, rich prairie lands of the Midwest. However, the first land-grant colleges stressed not farming but the classic curriculum of history, language and science that was considered the hallmark of an educated individual. Moreover, while Germany and

ARCHIBALD YELL
Governor of Arkansas
1840 to 1844

Gov. Archibald Yell was the first Arkansas governor to ask the state legislature to appropriate public funds to support agricultural research in 1842. (Photo courtesy Arkansas History Commission.)

Great Britain were laying the bases of scientific agriculture, Americans tended to view farming as a practical rather than a theoretical endeavor, and farmers disdained "book farming."

In the aftermath of the Civil War as the vast expanse of the Great Plains slowly came under cultivation, farmers in the settled Eastern states feared increasing competition. To deal with this anxiety, New York and New England farmers switched their production to fruits, vegetables and dairying. Continuous cropping of New England soils resulted in loss of soil fertility. In addition, farmers often believed they were purchasing adulterated or useless fertilizer; consequently, they demanded the employment of chemists to test and report fertilizer efficacy. With this purpose in mind, the skeptical farmers in New York and Connecticut agreed to support the first Experiment Stations.

The design of the state-based Agricultural Experiment Station utilized by the state land-grant institutions was laid out by Willard C. Flagg of Illinois Industrial University. Flagg envisaged researchers working in the laboratory and then conducting field tests to confirm their findings. As Flagg saw the Stations, each state would concentrate on its needs according to its unique climate, soil conditions and commercial agriculture. In 1881, teachers of agriculture presented a petition to Congress requesting that Experiment Stations be attached to land-grant institutions to scientifically deal with the present problems of the dairyman, the grain raiser, the stockman and the fruit grower.

A. W. Bishop, president of the Arkansas Industrial and Agricultural University (University of Arkansas) from 1873 to 1875, supported federal funding for agriculture. (Photo courtesy John Hugh Reynolds Photographs, Special Collections, University of Arkansas Libraries.)

The concept of federal financial support for the Experiment Stations came from Seaman A. Knapp, a professor of agriculture at Iowa State Agricultural College. In 1882 Knapp designed the legislative formula to ensure the establishment of Experiment Stations with an annual appropriation of $15,000 to each of the agricultural colleges. Realizing the potential for conflict between the academic and research functions of the University, Knapp constructed a triumvirate of the commissioner of agriculture, the college president and the professional superintendent to guide the Station's activities. The concept of cooperation foreseen by Knapp was not to be without the inevitable administrative conflicts necessitated by this unique approach.

In Arkansas, A.W. Bishop, the president of the newly established Arkansas Industrial and Agricultural University, had attended the meeting of the National Agricultural Congress of Educators in Atlanta in 1874 to discuss the possibility of federal funding for agriculture. The Arkansas land-grant institution was seriously under funded, and Bishop was eager to lobby Congress for additional money to fund agricultural instruction. From the University's founding in 1871, the curriculum had offered a course in agricultural science. However, the University had to abandon the course due to insufficient financial support and lack of interest. The location of Fayetteville in the northwest corner of the state was far from the center of cotton cultivation in eastern and southern Arkansas that dominated the state's economy.

Across Arkansas' state economy, the interest in agriculture was intense as falling cotton prices in the latter half of the 19th century caused farmers to forfeit their farmsteads to creditors. Farm protest movements such as the Farmer's Alliance and the *Arkansas Agricultural Wheel* newspaper, founded in Des Arc, Arkansas, hoped for a solution for the economic dilemma through political remedies, particularly currency manipulation, the free and unlimited coinage of silver to inflate the price of agricultural commodities. The *Agricultural Wheel* complained, "Farmers raise too much cotton and not enough grain and stock. The isolated farmer sinks under accumulated miseries."

The state legislature in 1886 responded to the desperation of impoverished farmers by proposing sweeping changes at the University, commencing scientific research in agriculture. The legislature enacted the Barker Act in March of 1887, reducing the president's salary to $2,000 a year and establishing the position of superintendents of agriculture and of mechanical arts at $1,600 each. Acknowl-

Depressed cotton prices in the 1880s dealt a serious blow to the state's economy, causing many farmers to forfeit their farms to creditors. (Photo courtesy Arkansas Department of Parks and Tourism.)

edging the failure of the University to teach agriculture-related courses, the University in 1887 appointed chemistry professor Albert Menke the first superintendent of agriculture and charged him with initiating a University experimental farm. Agricultural students who worked on the University farms were to be paid 10 cents an hour to be credited to bills for room and board. They would also have a tuition waiver and first preference to stay in the new dormitory, Buchanan Hall. The legislature appropriated $8,000 for the purchase of farm land and to pay for a course of study that would demonstrate methods of scientific farming to farmers. The generous tuition waiver in the act temporarily increased the number of students who attended the University.

The movement towards a state-based Agricultural Experiment Station system was given impetus by the political discontent of the nation's farmers. Agricultural protests created the political support necessary for the United States' Congress to enact the Hatch Act in 1887. The act appropriated $15,000 a year to support each state-university-based Agricultural Experiment Station.

Following the passage of the Hatch Act, Wilber Atwater, agricultural scientist; Samuel Johnson, the director of the Connecticut Agricultural Experiment Station; and George Cook, the director of the New Jersey Agricultural Experiment Station, drew up a 30-page report outlining the problems and "general character" of the new entity. First, Station administrations were to develop a method of

Albert E. Menke, a young chemistry professor, was appointed the first director of the Arkansas Agricultural Experiment Station in 1888. During his tenure as director, Menke started the experimental farm in Fayetteville and helped to organize the Stock Breeders Association and Farmers' Institutes. He established herds of Herefords and Holsteins and evaluated varieties of corn, cotton, wheat, grapes and strawberries. He also administered the fertilizer and seed testing program.

disseminating the research information already developed by European scientists, particularly from Germany, the locus of chemistry investigations. Second, agricultural questions to be dealt with were to be of "practical importance." American farmers would never approve of research efforts that did not have direct applicability to their immediate problems. Third, the most rigorous of scientific standards were always to be maintained. The scientific credibility of agricultural research was under close scrutiny from scientists in the basic sciences. Finally farmers had to be taught the results of the work and new approaches; the public relations aspect of the Experiment Stations was to play a vital role from its inception.

The Hatch Act was enthusiastically greeted by the Arkansas General Assembly, which officially accepted the $15,000 gift on February 17, 1888. Although only a small proportion of the money could go for buildings, the University used $4,000 of the first appropriations to build a brick building on the north side of campus and another $4,000 for scientific equipment. The need to deplete first year's appropriation to this extent reflected the University's penury.

The first director of the Arkansas Agricultural Experiment Station was Superintendent of Agriculture Albert Menke, who was considered the finest scientist on the faculty. On September 1, 1888, Menke officially took his position as the first direc-

Agriculture students who worked on the experimental farm were paid 10 cents an hour and given first preference to stay in the new dormitory, Buchanan Hall (top right). (Photos courtesy Photograph Collection, Special Collections, University of Arkansas Libraries.)

tor with a botanist, two chemists, a biologist, an entomologist, a horticulturist and a veterinarian on his staff. Menke reported on the pathetic condition of the proposed work. The prospective site for the Experiment Station's practical experiments consisted of "three acres in orchard, three acres in cultivation, forty acres untilled, a pair of antediluvian mules and a tumble-down shed." Menke requested an additional $20,000 for land, tools and stock and to build farm houses and facilities for dairy cattle. To deal with these difficulties, Menke conferred with an investigative state senate committee sent to Fayetteville to ascertain progress. Menke told Judge Lafayette Gregg that the Experiment Station needed $14,000 from the legislature. However, the Arkansas legislature was to provide niggardly financing for the first 35 years of the Experiment Station's existence. Instead, the Station depended on federal and limited University resources. To assist the Experiment Station, the White River Stock Breeder's Association, which Menke helped to create, purchased

Headquarters for the Experiment Station was build in 1888 at a cost of $4,000. The building was located near Maple Street behind the present Home Economics building. The building later was used for vocational teacher training, music, general extension and editorial services and was razed in 1972. (Photo courtesy Photograph Collection, Special Collections, University of Arkansas Libraries.)

Early research focused on practical problems facing stock raisers and fruit growers, such as pruning trees (shown above). One early Station Bulletin reported on the validity of folk beliefs, disproving the idea that dehorning cattle changed the proportion of fats and solids in milk.

land outside of Newport for grass and beef experiment purposes. Other sites in southwestern Arkansas were leased for experiments in cotton.

Menke realized the importance of combining scientific endeavor with practical result. He pledged the Experiment Station to deal exhaustively with problems of practical importance to farmers, stock raisers and fruit growers of the state. As was the case with European models, chemical testing of soils and fertilizers was to be emphasized. Unlike the Northern Stations that were interested in underlying scientific principles, Southern agriculture, still recovering from the devastation of the Civil War and a worldwide decline in staple prices, needed to devote itself to the practical bent that would deal with "remedial action" to bolster production of existing crops.

In 1886 the first Bulletin was published, a typewritten publication entitled *Experiments on Cotton and Corn in Drew County* by F.M. Bordeaux (Bulletin 1). This was followed by a single Bulletin the following year by George Purinton and by seven Bulletins in 1888.

The bulletins published in the second half of the decade reflected the need to move quickly on the multitude of problems that beset Arkansas farmers. In light of Louis Pasteur's discovery of the transmission of germs, scientists worldwide started exhaustive investigations into the nature of animal as well as human disease. Station research dealt with highly infectious hog cholera and Texas fever, the then-mysterious cattle affliction that occurred in the wake of the transportation of cattle from Texas. Other projects in the first years of operation included studies of sample fertilizers and fodder and studies involving cotton, corn, sorghum and tree diseases. The small staff began to grapple with the persistent problem of animal fodder since Arkansas had a deficit in satisfactory animal feed. Early bulletins dealt with the testing of the validity of folk beliefs, including a study disproving the idea that dehorning cattle changed the proportion of fats and solids in milk (Bulletin 5).

Experiment Stations throughout the country were under pressure to get immediate results. One farmer spoke for many when he stated, "We do not want science floating in the skies; we want to bring it down and hitch it to our plows." This utilitarian tradition permeated the first efforts of the Arkansas Station. An indication of the desire of the Arkansas Station to establish its credibility was the testing of commercial fertilizers by chemistry professor George Purinton, who emphasized that farmers should utilize their barnyard manure, an effective, low-cost option. In addition, Purinton concluded after experimentation that additives in the fertilizer such as superphosphate and ground bone as well as Pacific guano could not be economically justified given the low price of cotton. Purinton also commenced experiments in oats, corn and fruit growing.

Laboring under great difficulties, Menke and his staff achieved a prodigious amount of work. To assist in the work, given Fayetteville's isolation and climate, land was rented at Newport, Pine Bluff and Texarkana to experiment under the diversity of Arkansas soil and climate. Menke established a herd of Herefords and Holsteins as well as planting corn, wheat, potatoes and cotton.

Menke, like new Experiment Station directors in other states, was faced with the task of creating constituent support necessary for the perpetuation of the Station. Station scientists offered several programs to establish rapport with farmers, including analyzing farmers' fertilizer samples. Menke initiated public demonstrations called "farmers' institutes" to acquaint farmers first-hand with the work of the Station. These "farmers' institutes" later developed into the

Farmers' Institutes, which are now called Study Days, provided farmers the opportunity to see the latest developments in agriculture. The photo above depicts a farmers' day in 1925 in Fayetteville. (Photo courtesy Photograph Collection, Special Collections, University of Arkansas Libraries.)

Cooperative Extension Service demonstration or field day and the Experiment Station study day. Also, in an effort to directly assist the farmers, Charles W. Woodworth, the state entomologist, offered to identify any insect that was sent to the Experiment Station and forward suggestions as to control technology.

In 1888 the state legislature enacted a law requiring that all fertilizers offered in the state costing more than $10 be tested by the Experiment Station for a 15-cent fee with the results being attached to each package sold. The following year, corn and cottonseed varieties were also tested by Experiment Station scientists with the results being published.

The position of the Experiment Station under the aegis of the University created the problem of determining the teaching responsibility of Experiment Station scientists. To deal with this problem, a board of control was instituted consisting of the university president, the Experiment Station director and members of the board of trustees. Directors were to be given four-year contracts in recognition of the long-term nature of scientific studies.

Robert. L. Bennett, director of the Arkansas Agricultural Experiment Station from 1891 to 1903, was an advocate of agricultural diversification and alternative crops. He was concerned about soil erosion and encouraged planting of cover crops in the winter.

In order to firmly establish the Station, given the uncertain political atmosphere, Menke launched the Experiment Station scientists on a wide range of activities in addition to testing of fertilizers and seeds. New varieties of wheat, grapes, peas, strawberries, potatoes and sorghum were grown. The efficacy of cottonseed as an ingredient of compost was investigated as cotton growers tried to dispose of their surplus seed.

Charles W. Woodworth began a survey of plant pests and diseases in Arkansas. He had been trained under Professor Stephen A. Forbes at the University of Illinois. As a student he had imbibed his professor's philosophy on pest management, which has remained a hallmark of the Experiment Station till the present. Woodworth emphasized the importance of timely applications of poison, advising farmers to keep fields clean to prevent pests from taking control. Over application of insecticides, he said, had an adverse impact on crops. After becoming ill, Woodworth later left to serve in the California Experiment Station where he perpetuated his views on pest control.

In 1891, Menke left the Experiment Station to become the head of the faculty of pharmacy and later went on to a successful career in the steel industry. Menke was replaced as director by Robert L. Bennett, who had been in charge of the cattle-raising experiment in

Newport. Unlike his predecessor, Bennett, a native of Alabama, had practical farm experience. The new director established a branch location in Camden, Arkansas, for cotton studies.

Unfortunately, the deep agricultural depression of the 1890s resulted in a waning of enthusiasm for the Experiment Station. Currency manipulation rather than scientific farming was viewed by most farmers as the means to reverse the long-lasting decline in commodity prices. State financing was not forthcoming for agricultural experimentation. The number of students in agriculture virtually dried up, and Bennett dealt with a shrinking, demoralized staff. Bennett reported to the office of Experiment Stations that work was interrupted a great deal and much of the work begun was not completed due to staff departures.

By 1892, the Experiment Station was operating within severely constrained finances. The decline in cotton prices as well as the national industrial depression of the 1890s created financial pressures on the Experiment Station. In addition, the anti-intellectual attitudes of the Populists did not often lead to strong support for agricultural scientists. At the nadir of fortune of the new Experiment Station, the reduced staff of the Experiment Station consisted of Bennett as director and agriculturist; George Teller, chemist; R.R. Dinwiddie, veterinarian; J.T. Stinson, horticulturist; and two assistants.

To add to the discouragement of the staff, the spring of 1892 was cold and wet; consequently, the experiments planned for that year were not successful. As a result, the Experiment Station issued an informational bulletin on sorghum and sugar cane growing, neither likely to impact the vast majority of Arkansas farmers (Bulletin 22).

Given the state's climate, soil and economic parameters, alternative agriculture was not to be easily developed or implemented. The concept of diversified agricultural production, which was to be a hallmark of the Experiment Station, did not appeal to the farmers locked in the embrace of "King Cotton." To tenants and landowners alike, the maximizing of cotton production seemed to offer the only means of securing a profit for the year. Landowners had no incentive to encourage self-sufficiency among tenants, and tenants lacked the capital to implement the ideas of the Experiment Station. In 1900, one half of black tenant farmers did not own a cow. A tenant's average livestock was a mule, three hogs and six chickens. In perpetual debt to landlords and merchants and producing only a few bales of cotton, these farmers were bound to permanent poverty.

Chemistry professor George Purinton (top left) was one of the first Station researchers to evaluate commercial fertilizers. He supported the use of barnyard manure as a low-cost alternative.

Entomologist Charles W. Woodworth (top right) is considered the father of pest management in Arkansas. His guidelines for pest management, published in Station Bulletin 12, characterize the pest control philosophy used by many researchers today. Woodworth was considered the first trained agricultural research scientist employed by the Experiment Station. He later served as head of entomology at the University of California at Berkeley for 30 years. (Photo courtesy Bancroft Library, University of California, Berkeley.)

R.R. Dinwiddie, the Station's first veterinarian, helped develop a vaccine for swine cholera in 1902. (Photo courtesy Photograph Collection, Special Collections, University of Arkansas Libraries.)

Suggestions of the Experiment Station regarding alternative crops could not be readily implemented. A major portion of the agricultural population of Arkansas consisted of tenant farmers who rented land for a share of the crops. Cash-poor landowners were not prone to experiment with crops other than cotton. As Bennett pointed out, any alternative crop had to be a staple and in demand at all times and have a fixed and readily accessible market. As a result, farmers clung to cotton regardless of vagaries of price.

Bennett began Experiment Station demonstrations showing farmers that oats, wheat and grapes could be grown on worn-out cotton land and introduced the growing of grapes, plums, raspberries, strawberries and peas in northwest Arkansas. Bennett, like most agriculturists, believed that farmers should be urged to diversify given the fluctuations inherent in a one-crop economy. Bennett wrote that ". . . those who wish to realize much profit from their labor and money are compelled to abandon cotton and engage in other lines of farming." The Agricultural Experiment Station sought to work with "the most intelligent and progressive farmers in the community" to promote a diversified agriculture base. However, since cotton was the only negotiable commodity, the suggestion of agricultural scientists regarding crop diversification fell on deaf ears.

In 1896, the bottom of the depression, the Experiment Station issued its "live-at-home" guide to help tenants reduce their dependence on plantation commissaries that charged interest of upwards of 10%. Critics of existing farm practices preached the idea that disposable income could be expanded if farmers could be largely self-sufficient and limit their store purchases. However, in Arkansas, as in most of the South, sharecroppers and rental tenants were not free agents but had their efforts directed by the landowners. To deal with pandemic undernourishment among tenants, Bennett emphasized that farmers could augment their diet with Irish potatoes.

In the face of the apparent collapse of cotton prices, Bennett examined the possibility of economic rehabilitation of the rural economy by stock raising. In Bennett's analysis, farmers would have to switch to hogs and cattle and away from complete dependence on cotton. However, Bennett realistically appraised the immediate problems in raising livestock in Arkansas. The lack of veterinary service resulted in lackadaisical and ineffective combatting of cattle diseases. Better breeds of cattle would not be introduced until farmers ended open-range grazing and fenced their farmsteads; however, the practice of allowing cattle to forage for themselves proved difficult to halt. The farmers fed cottonseed meal and hulls to cattle, bringing about Experiment Station warnings that cattle could lose weight and die in Arkansas' warm and sultry weather (Bulletin 52). Finally, Texas fever discouraged the purchase of expensive, pure-blooded cattle that would sicken in Southern pastures.

Already by 1890, Arkansas farmers raised 99,689 hogs and 1,505,214 head of cattle a year. In addition, the manure served a useful purpose in helping improve worn cotton soils. However, an inadequate railroad system, distance from lucrative markets, insufficient corn and oat production and disease stifled the possibility of commercial livestock production without major improvement.

Another difficulty in creating a beef cattle industry in Arkansas was the lack of adequate forage. Substandard nutrition resulted in dying cattle and horses with the blind staggers, a nervous disorder caused by eating the staggerbush. The Experiment Station was to undertake long-term studies of optimal, nutritionally sound fodder as well as pasture improvement necessary to assist Arkansans in their endeavors to raise beef cattle. The Station also warned against use of spoiled corn as forage and advised that moldy hay or oats would cause an adverse reaction in animals.

In 1900, the Experiment Station started a search for suitable plants for summer hay pasture utilizing land at Newport furnished by the White River Valley Stock Breeders Association. The state's inconsistent weather patterns, punctuated by periodic summer droughts, complicated the problem of pasture improvement. As a result of preliminary work, Bennett concluded that red and crimson clover and alfalfa could be fed to piglets, sorghum and peanuts could be utilized for fodder, and a short period of corn fed prior to slaughter would be adequate to improve meat of hogs fed for slaughter. With beef, he suggested use of cottonseed supplemented with clover hay as a fodder (Bulletin 61).

A major part of the agricultural philosophy of Bennett and other soil scientists was the efficacy of legumes, especially cowpeas, to revitalize the increasingly eroded Arkansas farmsteads. The director pointed out that nitrogen-fixing legumes allowed for a 20 to 25% increase in yields in cotton and a 75% increase in corn. Cowpeas could be used as fodder as well as a method of crowding out noxious weeds from pastures.

In the search for diversification, C.L. Newman suggested that better varieties of wheat, such as commercial varieties of 'Tennessee Fultz', 'Indian Swamp', 'Purplestraw' and 'Red Wonder', be used

Wheat yields of 9 bu/acre in 1900 prompted Clifford L. Newman to evaluate varieties better adapted to Arkansas' growing conditions. (Photo courtesy Photograph Collection, Special Collections, University of Arkansas Libraries.)

since Arkansas production was below the national average. As a ground cover, soybeans and cowpeas and phosphate muriate of potash could be utilized. Given the lack of capital and the nature of Arkansas farm labor, farmers found it hard to follow these suggestions to improve wheat yields.

The one agricultural sector that afforded the greatest promise of tangible results at this time was horticulture. The northern counties of the state already had a flourishing truck farming industry, especially in apples. Unlike cotton experiments, fruit and vegetable experiments could be successfully conducted in Fayetteville. However, with capital expensive, it was difficult to convince farmers to expend much money on orchard improvements. The Experiment Station in 1897 issued a handbook for farmers on truck farming emphasizing the importance of rotation of crops and fertilizers. The handbook also suggested other crops that could be profitably grown, including asparagus and limabeans, cauliflower, carrots, collard, spinach, radishes, tomatoes and watercress.

Horticulturist J.T. Stinson endeavored to assist the growing number of truck farmers by writing a practical information bulletin to deal with the most common of insects injurious to fruits and vegetables, especially the codling moth (Bulletin 39). The Experiment Station emphasized that spraying Bordeaux mixture was essential in order to maintain healthy orchards. The expansion of the railroad lines into the Ozarks, especially into Washington and Benton counties, had afforded Arkansas producers entry into the Midwest market. Consequently, high quality produce was a necessary ingredient for successful sales.

In keeping with the practical nature of research, horticulturist J. T. Stinson in 1896 wrote a report dealing with the most common insects injurious to fruits and vegetables. The following year chemist George Teller published a basic handbook for farmers on fertilization and soil. (Photo courtesy Photograph Collection, Special Collections, University of Arkansas Libraries.)

Although the original intent of the framers of the Hatch Act was to emphasize original research as well as to deal with practical problems, financial exigency and inadequate facilities in Arkansas made basic research difficult. Nevertheless, chemist George Teller published his extensive analysis of the chemical properties of wheat and mill products (Bulletin 53), receiving favorable reception from the national director of Experiment Stations, Adam True. Teller also viewed himself as a disseminator of existing scientific information that would be useful to farmers. To help clarify the conflicting and often misleading claims of fertilizer advertisements, Teller wrote a basic handbook for farmers on fertilization and soil (Bulletin 47). He believed that this information would protect against fraud and wastefulness by giving farmers a basic working knowledge of the principles of fertilization.

A major concern of the Experiment Station was the question of erosion of the Ozark and Ouachita hill counties of northern and western Arkansas as well as drainage of the wet lands of the eastern and southern portions of the state. Bennett blamed farmers for their prevailing practice of not planting a winter cover crop or rotating crops. Moreover, Bennett regarded the gullying problems common throughout the state as a consequence of corn rows "not run on a level, but up hill and down hill with no idea of conformity to the slope of the field." Bennett, as well as most of his contemporaries, realized the difficult situation of changing the existing mode of agriculture tenure. Bennett viewed "farmers' institute" meetings as an essential element in effecting change by demonstrating tangible results.

Bennett established a Southern Branch Station at Camden to study the rapidly worsening problem of soil erosion in cotton land. The problem was compounded by the frequent moves of tenants and sharecroppers at the harvest, hence leaving the fields exposed and bare in the winter months, and the reluctance of farmers to make improvements. The planting of cover crops in the winter to be plowed under as "green manure" in the spring was difficult to implement. Bennett emphasized that planting oats, rye or crimson clover would prevent leaching of the soil. As a result, the farmers could buy the "cheapest and most durable" fertilizer, such as cottonseed meal or crushed cottonseed, that would suffice.

As a native Southerner familiar with Arkansas conditions, Bennett believed that Arkansas soils needed restoration and fertility preservation. However, the lack of adequate numbers of manure-producing animals combined with shallow plowing and up-and-down plowing on hillsides increased the rate of erosion. Bennett maintained that by planting cowpeas as a green manure, legumes could substitute for animal waste. In addition, cowpeas would serve as hay, helping Arkansans who usually had their cattle graze on the open range.

The scientists at the Experiment Station discussed the need for a change in attitude as a precondition to improved farming. At the turn of the century, life in Arkansas revolved around agriculture. Arkansas' capital--and largest city--had only 39,000 inhabitants, while most villages had a store, a blacksmith shop and a few basic businesses. More than 80% of Arkansans were involved in agriculture. Many were poor, living in ramshackle housing despite the presence of often-successful cattle plantations. The majority of the farmers ate salt pork, cornmeal, molasses made of sorghum and fruits and vegetables in season. Collard and turnip greens provided some variety as well as a source of vitamin B. This diet high in fat and carbohydrates and low in minerals, protein and vitamin B contributed to health problems, particularly pellagra caused by a lack of niacin. Arkansas' human as well as its financial capital had been depleted in the aftermath of the Civil War.

The high illiteracy rate led to suspicions about "book farming." The Experiment Station put out a variety of bulletins on practical dairying, livestock diseases, insects, the value of fruit and vegetable growing, soil analysis and forage crops to try to remove the barriers preventing diversification. Bennett continued his efforts to emphasize the use of cowpeas, "the poor man's fertilizer," which could increase soil fertility and foster greater productivity. He had hopes that practical demonstration would succeed where bulletins made little impact. "Farmers' institutes" were held to discuss crop rotation, animal and plant diseases, soil conservation, dairy and livestock husbandry. However, scientists recognized that the farmer's lack of capital resulted in his inability to deal with the decline in soil fertility caused by leaching and erosion.

In 1897 the Station abandoned work at Camden and Newport due to inadequate finances as well as the desire of the office of Experiment Stations to consolidate the limited funds on the Main Station. Bennett regretted the lack of state support and the fact that the land available for experiments was poor and not suitable for field work. He felt that the relative isolation of the Station in northwest Arkansas hampered its effectiveness statewide; in fact, he confided to Adam True, director of the office of Experiment Stations, that most farmers in the state were not aware that the Experiment Station existed.

In 1899, chemist G.L. Teller and horticulturist J.T. Stinson resigned. William G. Vincenheller replaced Stinson as an expert on fruits and vegetables as well as supervisor of the "farmers' institutes." Vincenheller believed that the "farmers' institutes" should concentrate on trial orchards and that only through personal contact and discussion could greater interest in scientific agriculture be generated.

To concentrate the Experiment Station's limited resources, the the "farmers' institutes" were discontinued in 1901. Vincenheller, who had been in charge of the institutes, found his salary reduced. He complained bitterly, but to no avail.

In 1902, Bennett commenced the long and highly productive relationship with farmers in the Grand Prairie area of Arkansas who were beginning to raise rice. Although for the first years the rice crop in Lonoke County was disappointing, the Experiment Station worked with 20 farmers on a few acres of leased land in Lonoke until 1910. In 1903, Experiment Station requested the assistance of the Department of Irrigation and Drainage in creating irrigation systems for rice. The Experiment Station utilized two varieties, one from Honduras and one from Japan. The work laid the foundation for the enormous expansion of rice cultivation on the Grand Prairie.

Agriculturist C.L. Newman was disturbed that wheat yield in Arkansas averaged only 8.9 bu/acre, 4 bu/acre below the national average. Starting in 1900, Newman tested 50 varieties of wheat to find

During the early 1900s, R.R. Dinwiddie (pictured here on left, with an assistant) vaccinated hogs for swine cholera on farms throughout Arkansas. The Station produced the vaccine in Little Rock. (Photo courtesy C.J. Brown.)

one adaptable to Arkansas conditions. Given the poor soil, he advised that growing cowpeas and soybeans as well as utilizing fertilizer such as phosphate potash would increase yields 56.5%.

The Arkansas Experiment Station served as a conduit of information regarding the advances in agricultural disease control from other state experiment stations. The knowledge garnered by scientists elsewhere transformed the approaches to disease. In 1882, Robert Koch isolated the microbe that caused tuberculosis. In 1889, Theobald Smith proved that ticks carried Texas cattle fever. In 1890, Koch devised the tuberculosis test for diagnosing bovine tuberculosis. In 1891, the federal government began free distribution of a vaccine for blackleg. In 1903, Marvin Dorset developed a vaccine for hog cholera.

The Experiment Station in its publications tried to inform farmers of these developments as well as the new methods of soil preparation for cotton and cowpeas. The Experiment Station's six scientists worked assiduously, supported by limited resources.

In 1903, worn out by the lack of institutional and state support, Bennett resigned as director. Bennett informed the head of the Office of Experiment Stations that the continual threat of political interference prevented scientists from adequately pursuing long-term projects. He resented the role of the Experiment Station's board of control in suggesting areas of scientific inquiry and annually reviewing work and was unhappy about the lack of interest in implementing Experiment Station suggestions. Bennett predicted that it would be difficult to secure a capable director for the Station unless politics was removed from the administration of the Station.

Bennett was succeeded by vice-director William Vincenheller, who was given the title of dean of the College of Agriculture as well

William Vincenheller, Station director from 1903 to 1908, promoted the organization of youth clubs to inform the youth about scientific farming. The clubs later became the genesis of the 4-H movement.

as director of the Agricultural Experiment Station. The appointment of Vincenheller came at an opportune time to win greater public acceptance for the Experiment Station with the implementation of the campaign to wipe out Texas fever in the wake of the discovery that tick eradication could eliminate the disease.

The working facilities in Fayetteville were improved after 1903. The state legislature finally appropriated some funds to assist the Experiment Station: $9,000 for an agricultural building and $5,000 for a dairy building and equipment.

The state legislature in the fall of 1903 appropriated $1,000 to assist veterinarian R.R. Dinwiddie, station pathologist and bacteriologist, in creating the program for tick eradication by systematic dipping of cattle. The Experiment Station moved cautiously in implementing tick elimination procedures that would entail the end of the open range. Farmers and livestock producers had long allowed their stock to roam free. Vincenheller castigated farmers who refused to cooperate in the dipping of cattle and the fencing of pastures, charging them with retarding the elimination of the Texas fever tick.

Dinwiddie also started working on a vaccine for swine cholera, and the Station produced the vaccine in Little Rock for distribution to farmers. Dinwiddie offered to personally visit farms if his professional expertise was needed.

In 1905 the legislature appropriated a $1,200 traveling fund for Dinwiddie, making it his duty to investigate outbreaks of disease and advise the owners as to the best means of dealing with them. The major diseases to be treated were glanders, Texas Fever, tuberculosis, anthrax, black-leg, hog cholera and rabies. In 1907 an additional $2,500 was allocated for tick eradication and enforcement of a livestock sanitary law.

The thrust of agricultural research changed as the deflation of basic commodity prices was reversed after 1900. In the first decade of the 20th century, the massive influx of Europeans into the United States contributed to an advance in agricultural prices. In addition, inexpensive oceanic shipping rates contributed to a worldwide market for agricultural commodities. The prices the farmer received for corn, wheat and cotton rose 36.6%, 49.6% and 58.7%, respectively, from 1900 to 1910. Agricultural scientists believed that better cultivation and seed, crop and land selection would permit the eventual doubling of yield. Adam True, the long-time head of the office of Experiment Stations, believed that greater support for research would eventually translate itself into increased staple production.

The year Vincenheller assumed his duties as director, interest in scientific farming got a major boost when the boll weevil started its depredation in the cotton South. Seaman A. Knapp, agriculturist and pioneer agronomist in rice cultivation, started to help Southern farmers deal with the boll weevil. Implementing his theory of demonstration to promote improved agricultural practices, Knapp estab-

The Roosevelt commission on country life helped strengthen interest in scientific agriculture. The first rural life conference (shown below on these two pages) was held in 1914 in Fayetteville.

lished the Farmers' Cooperative Demonstration. In 1905, J.A. Evans was appointed state agent to work with the Experiment Station to help gear up for boll weevil management.

To help diversify the state's agriculture and reduce its dependence on cotton, stock breeding and truck farming were considered alternatives to the dependence on cotton. However, progress was slow. Sharecroppers and tenants had little money to produce breeding stock. The scrub cattle in the South depended on natural grasses and forage, which were not nutritious. For better grasses, pastures had to be seeded and costly fertilizers utilized, neither of which farmers could afford. In addition, these farmers had to depend upon shippers to sell animals that could not be sold locally. The shippers bought for a low carload price, and farmers had to accept whatever prices were offered. To succeed in truck farming or fruit growing, the operator had to give close attention to proper cuttings, seed and cultivars as well as marketing. This type of farming required better management and knowledge than most farmers could muster. In addition, sharp price fluctuations, bad weather and dishonest commission merchants hurt even excellent farmers.

During Vincenheller's tenure as director, several researchers were employed, departments were reorganized, and facilities were improved. Charles Adams was charged with state inspection of nursery stock, plants and bulbs. V.A. Hooper, "an expert butter maker," was hired, and C.P. Norgord, a graduate of Wisconsin Agricultural College, was employed to commence efforts at breeding corn and cotton and to protect farmers from "unreliable and contaminated seed."

In 1906 Congress enacted the Adams Act to give the Experiment Stations additional funding to support basic research essential to major scientific breakthroughs. The act increased the appropriation

In 1905, the Station staff moved into a new facility. The building, which stands near Maple Street behind the current agriculture building, was later used for home economics and the University's infirmary and currently houses the department of biological and agricultural engineering. (Photo courtesy Photograph Collection, Special Collections, University of Arkansas Libraries.)

for the Experiment Station by $3,000 a year and raised it in increments of $2,000 a year until it was $30,000 a year. The following year the government enacted the Nelson Act, making the annual raises $5,000 a year with the target being raised to $50,000.

The Experiment Station was first organized into departments in 1906, including agronomy, horticulture, bacteriology and pathology, entomology, chemistry, animal husbandry, dairy husbandry and veterinary science. The department of plant pathology was set up in 1909. Home economics became a department in 1913 and was merged with agricultural chemistry in 1964. In 1921 the departments of animal husbandry and dairy husbandry became one department: animal husbandry and dairying. This was changed to animal industry in 1926 and to animal sciences in 1965 and to animal and poultry sciences in 1988. The department of veterinary science became the department of bacteriology and veterinary science in 1929 and merged with animal industry in 1950.

The completion of the Frisco Railroad provided new marketing opportunities for apple producers in northwest Arkansas. The apple packing scene above was taken around 1910 on Burch Grabill's farm, located one and one-half miles northwest of the University.

The photograph to the right shows William Vincenheller (right of center) at Arkansas' apple exhibit at the Louisiana Purchase Exposition and World's Fair in 1904 in St. Louis. (Photos courtesy Washington County Historical Society.)

Victor A. Hooper was hired as the Station's first expert butter maker in the early 1900s. (Photo courtesy Photograph Collection, Special Collection, University of Arkansas Libraries.)

The interest in scientific agriculture as part of the preservation of rural values underlay the Country Life Commission created in 1907 by President Theodore Roosevelt and headed by Liberty Hyde Bailey, agricultural scientist from Cornell University. The commission proposed a host of social and economic reforms to strengthen the efficiency of farmers as well as to preserve traditional agrarian society through education, particularly the formation of corn, pig, canning tomato and poultry clubs for boys and girls.

The Country Life Commission proposed that churches and schools had to combat rural isolation. In addition, improved public health facilities had to be the major objective of agricultural scientists. The building of farm-to-market roads was essential to allow the tapping of the agricultural hinterland. The Country Life Commission led to the passage of the Smith-Lever Act in 1914 for the formation of the Agricultural Extension Service. As the chairman of the County Life Commission, Bailey predicted that in the future, experts trained in every branch of agricultural science would help farmers to maximize production. Bailey's vision presaged modern scientific agricultural research.

In 1908, Vincenheller, disappointed by the lack of progress in implementing scientific agriculture and unhappy by his reduced compensation, resigned with complaints about allocations of funds and teaching demands on the faculty. Vincenheller felt that the

Charles Adams, a medical doctor and entomologist, was named Station director in 1908. He emphasized improved pest control practices for crops and livestock and was responsible for state inspection of nursery stock, plants and bulbs.

supervision of the Station by the board of control brought political considerations to scientific inquiry.

The new director, Charles Adams, had an M.D. degree from Kansas City Medical College as well as a masters degree from the University of Arkansas and a glowing endorsement from Adam True of the federal office of State Experiment Stations. He had joined the Experiment Station staff as an entomologist. In an overview of his new job, Adams viewed the heavy teaching load as his chief impediment to a successful research program and recommended that Station scientists teach only three hours each semester. Adams' plea went unheeded, and the Experiment Stations found it difficult to comply with the Federal mandate for limited teaching responsibilities. Consequently, the volume of work was reduced, and Adams found himself increasingly at odds with the University administration, which eventually forced him to resign in 1913.

During his five-year tenure, Adams emphasized insect and pest control. In 1912, the Experiment Station tested various sprayers to see if they provided a steady mist. Adams helped in the development of a hog cholera vaccine that provided a degree of protection from that dreaded animal malady. In particular, Adams emphasized that proper diagnosis was imperative to determine if swine had plague, pneumonia or cholera in order to better treat them. Until these diseases were controlled, the Yorkshire and Poland China hogs that

Adams wanted introduced into the state could not flourish. In addition, Adams continued to caution against the use of cottonseed as the major source of animal nutrition.

In 1912, dealing with prevention of plant diseases, the Station emphasized that a well-balanced fertilizer prevented twig and blossom blight. Adams estimated that at least one third of the apple crop was being destroyed. He estimated that 90% of the trees were falling prey to a host of diseases.

The proximity of orchards to the Experiment Station resulted in major efforts to upgrade their quality. With the building of the Frisco railroad in the early 1880s, farmers in the Arkansas uplands had access to Midwestern markets. In the next decade, Washington and Benton counties in northwest Arkansas became one of the leading apple-producing regions of the country. Unfortunately, poor cultivation practices, disease and insect infestation took an ever-increasing portion of the crop. Ernest Walker, the horticulturist, believed that the lack of basic maintenance was destroying the vitality of the fruit industry. The Experiment Station stressed that good drainage and fertilizer as well as pruning and spraying trees would improve the harvest. The underlying message of the Station was that successful farmers could actively work to improve crops rather than depending on divine providence.

The progress being achieved under Adams was confirmed by a visit from E.W. Allen, assistant national director of the office of the Experiment Station, who visited the Station in 1908. Allen was favorably impressed by the personnel and work and commented that the Station scientists were "attacking in an earnest and energetic way many of the chief problems of the state."

Even with improved physical facilities, Adams experienced difficulties with poor salaries and the interference of teaching and extension work. The failure of the legislature to allocate a yearly fund for support of ongoing projects disappointed Adams, who found it difficult to comply with federal guidelines on pure research due to the turnover in staff and the lack of a farm of sufficient size to perform field work.

Adams resented what he perceived to be political interference in the workings of the Experiment Station, and, like his predecessor, he disliked the role of the board of control. Adams believed that the problem of the Experiment Station was a lack of understanding and appreciation of its role in the promotion of economic growth.

The Experiment Station continued to work on tick eradication, hog cholera vaccine, control of San Jose Scale, codling moth, plum curculio and the cotton boll weevil. Martin Nelson commenced his rotation of corn, cotton and legumes. In addition, the Experiment Station tested various crops such as sorghum, kafir corn and velvet beans for their adaptability to Arkansas conditions. The department of animal husbandry explored the possibility of feeding rice bran and rice polish to pigs.

In 1913, Adams submitted his resignation. He had been unable to get sustained state funding for the Station. In addition, Adams opposed a legislative suggestion to either move the Station to Little Rock or share the Hatch Act money with other academic institutions.

Martin Nelson was appointed Adam's successor and directed the major investigations into apple and peach tree borers, continued fertilizer tests and experiments in crossbreeding of cotton. Nelson reorganized the Experiment Station to relieve himself of direct investigation. The Experiment Station established a plant for the production of hog-cholera serum in Little Rock. Nelson believed these activities would demonstrate the vital role that the Experiment Station could play in the improvement of farming. His optimism was to be unfounded, as Governor George Donaghey refused to allocate any money for the Experiment Station. As a result, Nelson curtailed cotton experiments. Nelson felt that without state support, the Experiment Station could not fulfill its role as a catalyst for change.

University of Arkansas President John Hugh Reynolds realized that the concept of higher education and scientific agriculture re-

Martin Nelson served as dean and director from 1913 to 1920, vice director in charge of research from 1920 to 1928 and head of the department of agronomy until 1940. Nelson was a strong advocate of small grains, oats and wheat. In 1982, Station researchers released a wheat variety in his name.

quired a major selling job to Arkansas citizens. On January 13, 1913, Reynolds delivered a speech to the Arkansas State Horticultural Society at Fort Smith entitled "Thirty-Three Million Acres and the University." The thrust of his argument was that the failure of the legislature to pass a special tax for the University was undermining economic progress. He referred to the resignation of Charles Adams and others as a consequence of low salaries and inadequate support and asserted that this trend would undermine economic and civic progress. For the last six years, Arkansas had been spending about eight cents per person, or forty cents for an average family, on the university, the price of a dozen eggs.

The creation of the Agricultural Extension Service in 1914 by the Smith-Lever Act meant a major step forward in the promotion of scientific agriculture. The *Arkansas Gazette* at Little Rock had long decried the fact that, with most Arkansas farmers being illiterate, Experiment Station bulletins were too difficult to read and comprehend. Demonstration projects could provide tangible evidence of the efficacy of scientific agriculture.

Nelson was fortunate in that the establishment of the Cooperative Extension Service provided a conduit for the application of basic work done at the University. Moreover, with the boll weevil devastation in southern Arkansas, farmers were far more interested than they had been in the basic work of pest control, stock raising and crop rotation. Moreover, the collapse of cotton prices in the fall of 1914 reinforced the long-standing appeal for diversification.

County agents were viewed with a jaundiced eye by skeptical farmers who disregarded the first extension agents' advice. They resented the idea that an outsider could question--or even had the right to question--their knowledge of how to farm and cultivate crops. Seaman Knapp believed that the demonstration farm would serve as the necessary bridge between book and practical farming. Knapp thought that the message of the agronomist could be embodied in a few cultivation practices. Farmers should be taught how to rotate crops of cotton, corn and cowpeas and how to use manure and commercial fertilizers. Farmers had to be familiarized with the principles of livestock production and pasture management and to be informed of new labor-saving devices. Knapp told the agents not to patronize farmers or denigrate the traditional farming methods but to integrate their folk beliefs with the new, scientifically based methods being developed by the Experiment Station.

Nelson's request for addition funds for the Experiment Station was passed by the Legislature but vetoed by Governor George W.

Hays. In 1916, the state legislature voted the first permanent appropriation to support ongoing projects. Finally in 1916, the Arkansas legislature passed a biennial appropriation of $49,000 supported by a millage tax. The money would be used to lease land for a cotton substation. Although the money was appreciated, Nelson had hoped for substantially higher funding levels to achieve the goals of the Station. He complained that the laboratories needed to be enlarged by at least five times. Small laboratories and cramped quarters resulted in difficulty meeting the demands placed on the Station. Nelson promised that, if given reasonable funds, the station would repay many-fold the value of its appropriation. He dismissed charges that the Experiment Station aided only the affluent farmers and maintained that the work of the Experiment Station moved the entire state's agriculture ahead.

In 1915, the Experiment Station worked with the state veterinarian to create tick eradication districts. The Station estimated that Arkansas had some 236,000 infested cattle with lessened milk flow, decreased market value, diseases and worthless hides. When a county agreed on a district, 25 to 40 concrete dipping vats were built so that farmers would not be more than three miles from a vat. Cattle were dipped every three weeks in summer in a solution of white arsenic, sodium carbonate and pine tar boiled in water. The program ran into some resistance as a result of a natural suspicion of an arsenic mixture, and some cattle owners opposed cooperation in the program. The Experiment Station estimated that the state

Tick eradication was one of the Station's most successful research programs from 1906 to 1919. The Station published Bulletins 90, 93, 101 and 119 on tick eradication and Bulletins 132 and 160 on related laws and regulations. This photo shows an early tick eradication camp site. (Photo courtesy Washington County Historical Society.)

would save some $5 million, or about $8.00 per head of cattle, when Arkansas adopted this program.

Because of the tick problem, Arkansas cattle could not be bought as dressed beef steers. The state was forbidden to ship live cattle from tick-infested areas across tick-free areas except for immediate slaughter unless a federal supervisor verified that the cattle had been dipped twice in 10 days. Stock raisers were unable to import pure bred cattle to Arkansas because a large number would die of Texas fever. Newton County was the first county to adopt a tick dipping program. In 1917, the state legislature allocated $50,000 to systematically wipe out ticks from western Arkansas, and northwestern Arkansas was declared tick free.

A major agricultural success story was occurring in rice. The substation in Lonoke was active in the exploration of improved rice production and effective use of rice byproducts. By 1913, Arkansas produced 4.8 million bu of rice/year, approximately 17% of the nation's crop. In addition, the Experiment Station tackled the problem of the use of rice hulls and bran, the byproducts of milling. In 1915, Arkansas was producing 43,560 tons of rice bran and 14,520 tons of rice polish annually, and the war demand for corn and wheat accelerated the need for these rich byproducts. However, rice bran became rancid and pest infested in the summer, and a rice-bran diet produced a soft, oily fat on livestock that packers did not desire. Consequently, rice bran was used only as a supplement to corn or rice polish.

Martin Nelson was particularly interested in production of oats, the third largest crop acreage by 1918. Martin pushed for experiments with different varieties, for oats were relatively free from insect pests and disease. Production was concentrated in the Ozarks area. As Nelson said, the value of the oats was low due to poor culture, poor seed and less fertile land. In the Ozarks, 'Red Rust' oats were planted in the spring and winter or in the fall, and the crop was cut for hay. In 1907, the Experiment Station had leased two farms, and between 1909 and 1913, various varieties of oats were tested. Nelson experimented to determine optimal planting dates for oats to avoid winter kill due to early frosts.

The Experiment Station was always interested in agricultural diversification. One crop already in production that appeared to be suitable for extensive and rapid development was sweet potato, which scientist W.H. Wicks estimated could make $70/acre. By 1910 Arkansas grew 1.6 million bu, mainly for consumption as animal feed. The Arkansas sweet potato growers needed to establish standards of grading, insuring that the potatoes would be "attractive . . . free of disease, bright, and well cured." One of the essential ingredients for successful farming was the need for a cooperative marketing organization, difficult in a state segmented by race and characterized by highly individualistic ethos.

In 1916, W.H. Wicks, working with the State Sweet Potato Growers League, published a bulletin summarizing the available information on sweet potato cultivation (Bulletin 124). The state had been a major producer of sweet potatoes in the nation with 4,685,308 bu/year, but growers needed markets outside Arkansas. By fostering sales in the months after the New Jersey sweet potatoes had been consumed, Arkansas growers could maximize profitability. However, even with cooperatives, Wicks warned that the initial outlay to plant sweet potatoes would be beyond the means of many Arkansas farmers.

The impact of World War I and the accompanying high wheat prices renewed interest in wheat. The Experiment Station had already conducted wheat experiments from 1897 until 1900 but had attracted little attention. In 1907 the Experiment Station commenced variety tests. However, farmers in the Ozarks with poor seed, eroded soil, no modern equipment and inadequate storage saw yields of only 11.9 bu/acre, far below the national average. The Station started wheat experiments dealing with the difficult agricultural conditions of the hill countries. In particular, Nelson was convinced of the need to change the emphasis on devoting seven-eights of all acreage in Arkansas to corn and cotton, which was robbing the soil of vital nutrients. Winter wheat could protect against erosion, and the wheat could be utilized for flour.

Arkansas was a major producer of sweet potatoes in the nation, and W. H. Wicks wrote Station Bulletin 124 to help promote the production of sweet potatoes as an alternative crop.

During the shortages of World War I, the bulletins on wheat and potatoes were the most desired by Arkansans with 40,000 copies printed in contrast to the usual runs of 5,000. With experiments on varieties of Irish potatoes, the Experiment Station advocated a change from the commonly used Bliss Triumph to the Irish Cobbler, Hebron and Early Ohio. In addition, the Experiment Station reported the possibility of growing a second crop of potatoes for the following season's seed in the Arkansas River region.

In 1917, the state legislature passed an act that established the Arkansas State Plant Board. The law decreed that the Agricultural Experiment Station would be represented on the board by the heads of the departments of plant pathology and entomology. George Becker, who was with the Experiment Station from 1910 and head of the entomology department in 1912, designed the law. He was designated state entomologist and travelled throughout the state to inspect nurseries. Becker resigned in 1918 to enter the military service. In 1919 when he returned, Becker was employed by the board as chief inspector until 1926 when he resigned to work with the Federal Horticultural Board and later with the Foreign Plant Quarantine Service.

In the aftermath of the outbreak of World War I, the demand for agricultural products surged. Under Martin Nelson's leadership, the Experiment Station built a demonstration silo. Plant pathologists surveyed the entire range of plant diseases in the state. Agronomists

George Becker, head of the department of entomology, helped write the law that created the Arkansas State Plant Board in 1917. This organization, linking Station research, extension and the state, became a model for other states to follow.

worked on a soil survey as well as tests of forage crops, wheat, rye, oats and barley.

The increase in funding after 1916 allowed the Experiment Station to purchase improved scientific equipment to begin studies of magnesium, sulfur and phosphoric acid. In addition, the Experiment Station initiated studies of poultry for egg production. Finally, as a consequence of Martin Nelson's studies on crop rotation, the Experiment Station was able to offer advice to farmers on methods to restore worn-out soils.

The Experiment Station's first 30 years had been difficult due to inadequate funding and a reluctance of farmers to accept the suggestions of the Experiment Station. However, the outbreak of World War I and high crop prices seemed to offer a far greater opportunity. The war also created a public relations challenge to the very existence of the Experiment Station on the Fayetteville campus.

The College of Agriculture received its first appropriation for a soil survey in 1917. This money laid the basis of what was to become a detailed survey of the soil chemistry throughout the state. This photo shows Station researchers and students using a soil probe to measure soil water content by weight.

Chapter II

The Experiment Station and State Politics 1914-1923

Agriculture in Arkansas in the second decade of the 20th century still revolved around cotton. Many farmers were poor, illiterate tenants with little knowledge of or interest in scientific farming. An estimated one third of the tenants moved every year; consequently, landowners were reluctant to invest in major capital improvements. Nutritional deficiencies and diseases such as tuberculosis and malaria sapped the physical vitality of many farmers. The isolated mountaineers in the Ozark and Ouachita Mountains eked out a living on eroded hill tracts. Bereft of money, these sturdy farmers raised corn, hogs and cattle while often trying to supplement income through seasonal work in the coal mines of the Arkansas Valley, the sawmills, the wheat harvest or the fruit harvest.

The Main Experiment Station, located in Fayetteville in the hills of northwest Arkansas, was regarded as relatively inaccessible by the farmers of the state, and the Station had a difficult time impacting agricultural practices. In addition, Station bulletins were considered abstruse and unrelated to immediate problems. These attitudes made the Station vulnerable to political attack.

The person who was to spearhead the attempt to move the Experiment Station out of Fayetteville was a transplant from Indiana, E.J. Bodman. Bodman had arrived in Arkansas in 1904 to work for the Union Trust Company of Little Rock. Frustrated by the lack of progress in agriculture compared to that in his native Indiana, Bodman announced the formation of the Profitable Farming Bureau in 1914 to get public support behind the Experiment Station and the newly created Extension Service.

Bodman went to Fayetteville to speak to Martin Nelson and University President John C. Futrall about his ideas and concluded that the legislature must appropriate a half a million dollars to give scientists adequate facilities. In order to generate public support, Bodman convinced Fred Heiskell, editor of the highly influential

E. J. Bodman, vice president of the Union and Mercantile Trust Company in Little Rock, was a strong supporter of agricultural research but publicly criticized the Station and the University for not being aggressive in helping Arkansas farmers. He established the Profitable Farming Bureau of the Little Rock Board of Commerce, which provided a platform to express his views on agriculture. Bodman's continuous criticism created years of political turmoil at the Station. (Photo courtesy Special Collections, University of Arkansas Libraries.)

Arkansas Gazette at Little Rock, to start highlighting information on scientific agriculture. Heiskell hired Bert Riley to go to Fayetteville and work with the Experiment Station to produce popular items to run in the paper. Although skeptical about the efficacy of this strategy, Futrall agreed to work with Bodman.

In 1917, America's entry into World War I imposed additional difficulty on agricultural research; the impact of inflation on labor and the cost of printing resulted in pressure on the slim Experiment Station budget. Higher prices and wages resulted in a loss of personnel to other positions. In addition, some scientists volunteered for the Army. The Station was starved for funds in spite of its potential to contribute to the war effort. In particular, the Experiment Station needed a large, modern facility essential for conducting experiments.

To augment their efforts, the Experiment Station cooperated with the Monticello District Agricultural School and Magnolia Agricultural School. Station scientists tested wheat varieties in view of massive demand both in America and from the Allies and tested Irish potato varieties due to the dramatic increases in food prices. Cottonseed meal and rice bran were investigated as animal feed substitutes. Entomologists studied pests, including the apple wooly aphid, the peach tree borer and the round-headed apple tree borer. The Experiment Station cooperated with the United States Army to

In 1918 H. M. Cottrell from Memphis was appointed the agriculturalist for Bodman's Profitable Farming Bureau. Cottrell recommended farming practices that he said would increase farmer's income. The Station later challenged Cottrell's recommendations. (Photo courtesy Special Collections, University of Arkansas Libraries.)

test water quality at temporary army bases as well as to assist in control of animal diseases; however, the absence of a Station veterinarian inhibited this work.

In June of 1918, Bodman's Profitable Farming Bureau organized a trip to examine facilities at the land-grant institutions in Missouri, Iowa and Wisconsin. Governor Charles H. Brough, former chairman of the University of Arkansas economics department, joined the group for the trips in Wisconsin and Iowa. The *Arkansas Gazette* gave heavy coverage to the trip and to Bodman's role in promoting agriculture. The Little Rock Board of Commerce published the findings contrasting the situation in Arkansas with that at the well-endowed facilities at other land-grant institutions.

Disappointed at the lack of response from the legislature, Bodman announced in December 1918 that H.M. Cottrell, an agricultural expert employed by the Memphis Bureau of Farm Development, would become the agriculturalist for the Arkansas Profitable Farming Bureau. With cotton prices reaching an all-time high, Cottrell seemed the harbinger of a promising future. Cottrell officially took his position in January 1919 after publicly rejecting an offer from the Memphis Bureau of Farm Development to top his Arkansas salary by $1,500. In his first published statements, Cottrell outlined his plans to raise agricultural production in the state by $300 million. He recommended the utility of sweet potatoes as hog feed in areas unsuitable for corn and suggested an emphasis on poultry and egg production. He advised the planting of velvet bean between rows of corn to restore soil fertility and predicted that use of legumes could double cotton production within five years. Cottrell also promised that farmers who adapted his proposals could triple their income. By April 1919, Bodman suggested that Cottrell should be appointed director of Extension in order to bring dynamism to scientific agriculture. Arkansas agricultural experts welcomed Cottrell as an ally but soon realized that his personal aspirations threatened them.

The immediate post-war prosperity and the constant drumbeat of publicity finally produced results. In 1919 the University received funds to purchase 423 acres for $123 an acre to be used for fertilizer experiments as well as to promote ongoing projects. Fifty acres were rented at Scott, and field experiments were carried on in 41 counties. Martin Nelson, in a bulletin entitled *Nature and Progress of the Work of the Agricultural Experiment Station* (Bulletin 158), tried to show the workings of the Experiment Stations and facilities. Nelson articulated the Experiment Station's philosophy that improved breeds of livestock, diversification of crops and use of cover crops would restore the productivity of the agricultural sector.

The Station's 423-acre farm in Fayetteville was a showcase for farmers to visit and observe new developments in agricultural research. (Photo courtesy Special Collections, University of Arkansas Libraries.)

Despite efforts at cooperation, relations between Bodman and University President Futrall deteriorated as Futrall resented the intrusions on University business by a private organization, particularly given the University's allocation commitment to purchase the farm. On the other hand, Bodman believed that Futrall was not willing to push for a first-class agricultural school comparable to the institutions in the Midwest.

In June 1919, Bodman announced another publicity tour of Kentucky, Indiana, Michigan and Illinois. The accompanying press releases emphasized the well-endowed Agricultural Experiment Stations of these states. On June 16, 1919, Bodman gave an interview at Indiana University to the accompanying *Arkansas Gazette* reporter stating his disappointment with Arkansas' efforts. Bodman told the newspaper the underfunded facilities in Fayetteville could be augmented by a successful publicity campaign aimed at businessmen as well as farmers.

At the conclusion of the trip, Bodman called for drafting of a proposal to present to the University of Arkansas board of trustees. The proposal attacked the Experiment Station, the University and Extension and claimed that the Experiment Station was not aggressive in promotion of agriculture. The proposal was forwarded to Governor Charles Brough, chairman of the board of trustees.

On June 21, 1919, Governor Charles Brough convened a special meeting of the board of trustees to discuss Bodman's charges.

Bodman presented his allegations, backed by Cottrell and *Arkansas Gazette* editor Heiskell. Futrall appeared to refute the charges accompanied by several witnesses, ironically not including Station director Martin Nelson. C.W. Watson, the assistant director of Extension, challenged the credentials of Cottrell, ridiculing the idea of planting spanish peanuts instead of corn for hog feed, and pointed out that planting velvet beans between corn rows was unproven and expensive even if sufficient supply were available. Watson pointed out that the instructions of the Arkansas Profitable Farming Bureau on egg preservation were incorrect. The acrimonious debate resulted in charge and countercharge.

After the presentations, Bodman and Heiskell were convinced of the hostility of the University and the agricultural establishment. The two men stormed out of the meeting vowing to bring their case before the public. The board expressed its support of Futrall and condemned Bodman. On June 25, Heiskell attacked the board's decision and proposed a three-man committee to investigate the charges, but the University rejected this suggestion.

Futrall realized that Martin Nelson, a prodigious and assiduous researcher, did not have a flair for public relations. E.W. Allen, chief of the office of Experiment Stations, became involved in the search for a successor for Nelson. As Allen viewed it, the state legislature's practice of allocating funds and not appropriating the entire sum had become a pernicious practice that undermined morale at the Station. Allen believed that, due to its relatively isolated location, the Station had not made a very strong impression on the state as a whole and felt that the University of Arkansas had not backed the Experiment Station. As a result of the unfavorable publicity, Allen pessimistically thought the future of the University and the Experiment Station was in serious jeopardy due to insufficiency of vigor in presenting their case. Although impressed by both Futrall and Nelson, he believed that the political opposition in Little Rock had tried to discredit the Experiment Station's work.

In 1920, Bradford Knapp, the son of Seaman Knapp, the head of the farmer's cooperative demonstration work, was named director of the Experiment Station and dean of the College of Agriculture with Nelson becoming vice-director. The legislature in 1921 appropriated a higher millage tax for the University, increasing it from four-ninths of a mill to one mill as well as granting $15,000 to the Experiment Station for a soil survey. Moreover, $50,000 was allocated for the Experiment Station for a 500-acre tract in central Arkansas. To

In an attempt to gain public support for the Station, Bradford Knapp succeeded the more reserved Martin Nelson as director in 1920. Knapp led the fight to keep the College of Agriculture and the Station in Fayetteville. However, the struggle eventually lead to Knapp's resignation four years later.

strengthen the Station, Knapp believed that salaries of the scientists needed to be raised 30%. To accomplish this objective, state appropriations needed to be increased between 75 and 100%.

The travails of the Experiment Stations were not over. In 1920, the price of cotton fell 50%. The economy of the South was devastated with the end-of-war-time prosperity, and the agricultural depression deepened. As Bradford Knapp viewed it, "There was never a time when farmers felt more keenly the need of the work of the Experiment Station." In 1921 businessmen who favored moving the University of Arkansas to Little Rock pressured the legislature to enact a measure realigning higher education in Arkansas. The proposal sent a shiver of fear down the back of Knapp, who feared the disruption of experiments as well as the accompanying costs of dislocation. In 1922, Governor Thomas C. McRae endorsed a proposal that the state purchase Camp Pike outside of Little Rock and relocate the Colleges of Agriculture and Engineering there. However, General John J. Pershing vetoed the conversion of the post to a new site for the University of Arkansas.

Bradford Knapp countered the relocation idea by maintaining that the Experiment Station was involved in an ever-increasing range of activities. Moreover, he contended, the operating cost of the Experiment Station was low. Knapp believed that a system of branch stations would be too much of a burden for the staff. On the other hand, Knapp realized that the Station would have to convince big, successful farmers of the efficacy of the Station.

To accomplish these objectives, Bradford Knapp made several changes in personnel and research. Knapp brought on staff Barnett Sure to study vitamins contained in common foods. Sure, an assiduous researcher, was to become a co-discoverer of Vitamin E as well as a national authority on vitamin deficiencies. Sure first tested the velvet bean that had been emphasized by Cottrell and concluded that "velvet beans should not be fed to animals . . . to obtain the best growth."

Martin Nelson was placed in charge of the fertilizer experiments on the new University farm, beginning what was to become three decades of experiments on crop rotation and fertilizer applications adjusted to soil type. In agronomy, J.O. Ware commenced his investigations of crossbreeding of cotton hybrids and optimal planting practices of cotton. L.W. Osborn, working with the Arkansas Seed Growers Association, experimented with new varieties of sorghum, clover, rye and oats. With the help of a small legislative appropriation, the staff completed the first comprehensive soil survey for Arkansas. S.R. Stout sponsored the annual Arkansas State Egg Laying Contests to secure superior breeding stock. For dairying experiments, the University acquired Jerseys, Holsteins and Ayrshires.

Dwight Isely, hired as an entomologist, began his long career in pest management with a mixture of biological controls as well as

In the early 1920s, Gov. Thomas McRae supported a proposal to move the Colleges of Agriculture and Engineering to Little Rock, and the Arkansas General Assembly debated a proposal to move the entire University to Pulaski County and the Colleges of Agriculture and Engineering to Russellville. (Photo courtesy Arkansas History Commission.)

chemical pesticides. Isely graduated from Fairmont College (Wichita State University) and received a masters degree from the University of Kansas. After a year at Cornell University, he joined the U.S. Department of Agriculture. In 1921 he joined the faculty at the University of Arkansas where he served for the next 35 years. Isely was to develop the technique of determining the economic threshold for application of pesticides by scouting fields to statistically measure the population levels and was to train a generation of entomologists in his methods.

Agronomist H.R. Rosen concentrated his work on crossbreeding of oats and of wheat. Entomologist W.J. Baerg studied poisonous arthropods leading to a self-experiment on his own physiological reaction to the bite of a black widow spider (Bulletin 325). Rural economist A.D. McNair, an expert farm manager, gave farmers accurate cost estimates on what could reasonably be expected from various crops. Isely and Baerg published a bulletin in plain language to explain what farmers could do about control of boll weevils by a combination of biological control and calcium arsenate (Bulletin 190).

Despite the efforts of agricultural scientists, an internal structural problem of Arkansas agriculture--the prevalence of tenancy--precluded major advances in agriculture. In the counties of eastern Arkansas, over two-thirds of the farmers were either share or rental tenants. Sharecroppers furnished labor, and the landowner pro-

Station researcher Barnett Sure was an international authority in nutrition and vitamins and co-discover of vitamin E. (Photo courtesy University of Wisconsin - Madison #x252231.)

S. R. Stout (top left), one of the early researchers in poultry, helped manage the annual Arkansas State Egg Laying Contests to secure superior breeding stock.

Dwight Isely (top right), a pioneer in pest management, followed closely the pest control strategies of Charles Woodworth. Isely's philosophy for refining insect control programs was based on the biology and behavior of pests.

W. J. Baerg's research in Panama, Mexico and Jamaica on poisonous arthropods, such as spiders, tarantulas and scorpions, brought him international recognition. In one experiment, Baerg (bottom left) checked himself into a Fayetteville hospital and induced a black widow spider to bite him to study the nature of the resulting symptoms.

vided the land, house, animals, equipment and a small cash advance for making the crop and for food for the sharecropper's family. The tenant received half the gross proceeds of the cotton crop minus deduction for expenses. The dramatic fall in cotton prices in 1920 and 1921 meant that the sharecropper's expenses outstripped his share of the crop. Due to inadequate sources of credit, banks forced landlords to pay 16% interest and tenants as much as 40%.

The low purchasing power of tenants combined with their frequent relocations weakened all of Arkansas' rural social institutions, including churches, schools, libraries and cooperatives. Individual farmers were impoverished and had poor diets and inadequate medical attention. Given the rural situation, it is not surprising that the scientists working at the Experiment Station felt their efforts could not be appreciated.

In view of these difficulties, the Experiment Station's political support appeared shaky. The 1923 session of the General Assembly debated a proposal to move the entire University of Arkansas to Pulaski County with the Colleges of Agriculture and Engineering moving to Russellville where the District Agriculture College was located. To reply to accusations as to the contribution of the Experiment Station to the state, Bradford Knapp issued a bulletin entitled *Facts About College of Agriculture* (Bulletin 182) to discern the real state of agricultural research. Knapp emphasized that 1,000 to 1,400 farmers had taken short courses during Farmer's Week and maintained that Station scientists were knowledgeable about the practical aspects of farming and grew some $40,000 a year in produce. With costs being a "mere pittance," the Experiment Station was working on plant breeding, insect control, investigations of methods of storage, processing, keeping, marketing, shipping and handling farm products. Knapp contested newspaper stories about the staff's credentials, maintaining that half of the staff had been awarded masters degrees and four had doctorates. As to their personal experience, Knapp countered that all but two had either been born on a farm or done farm work. Knapp concluded that the scientists were men of the highest moral character and churchgoers and worked as volunteers for charitable institutions.

In addition to this spirited defense, Knapp, President John C. Futrall and Vol Walker, former president of the Arkansas Bar Association, invited members of the legislature to visit both Russellville and Fayetteville to inspect the facilities. The legislators heard a fervent presentation by both Futrall and Knapp on the importance of agriculture to the University, and agricultural students presented a petition supporting the location at Fayetteville.

Bradford Knapp, disillusioned by the fight, resigned as director in 1924 to take a job as president of Oklahoma Agricultural University. Finding a successor to Bradford Knapp was not easy due to the political controversy surrounding the Station and the need for a candidate of excellent administrative and educational skills as well as political sophistication. The federal office of State Experiment Stations once again had to assist in the selection.

Dan T. Gray was selected in 1924 to serve as dean of the College of Agriculture as well as director of the Experiment Station, positions he held until 1939. A native of Missouri, Gray was educated in Missouri and at the University of Illinois before receiving an appointment to be professor of animal industry at Alabama Polytechnic Institute for eight years and a professor at North Carolina State University for seven years. He had made an outstanding reputation as secretary of the Association of Southern Agricultural Workers from 1913 to 1920 and as president from 1920 to 1921. In 1920 he was named dean of the School of Agriculture in Alabama. His appointment as dean of the college of Agriculture and director of the

Plant pathologist H. R. Rosen conducted crossbreeding research with oats and with wheat. In 1979, the Station released a variety of wheat named in his honor.

Agricultural Experiment Station in Arkansas in 1924 was hailed as a major addition. Gray had considerable experience dealing with problems of livestock in the South and also was to be the animal industry editor for the widely read *Southern Ruralist*. He was sensitive to the questions of the federal role in agriculture after having served as the administrator in charge of the federal government's livestock extension program during World War I.

Despite efforts to satisfy critics of the University and the Experiment Station, advocates of the relocation still refused to be reconciled to the Fayetteville location. On February 8, 1923, an open meeting was held in Little Rock to discuss the proposed relocation of the Experiment Station

In the legislature, the Senate Education Committee supported the removal bill with a "do pass" recommendation while the House Education Committee registered its strong disapproval. Under heavy lobbying by the University's friends, the Arkansas State Senate voted to indefinitely postpone further consideration of relocation. Although the Arkansas University Removal Association, which had initially proposed the measure, threatened to continue the struggle, the University and the Experiment Station were to remain in Fayetteville.

Dan T. Gray, dean of the school of agriculture in Alabama, was named dean of the college and Station director. He served the longest tenure as director, from 1924 to 1939.

The prolonged struggle ultimately was decided to the benefit of the Experiment Station; with the question of the location finally resolved, the legislature was to approve the building of a modern facility for the College of Agriculture and the Experiment Station. In addition, adequate funding would finally be made available for branch stations throughout the state in order to deal with the specific problems faced by the various sectors of Arkansas agriculture.

Under the leadership of Dan Gray and with increased federal as well as state money, the Experiment Station personnel could finally turn their full attention to the problems and opportunities confronting Arkansas farmers.

College of Agriculture, faculty and staff, spring 1924: front row (l to r), Carrie Plunkett, Lolita Brogdon, Virginia Wardrip, Miss Craig, June Letsch, Madge Johnson; second row, C.K. McClelland, A.D. McNair, T.A. Fritts, J.R. Cooper, Dan T. Gray, Mrs. Brandon; third row, H.E. Dvorachek, V.H. Young, John W. Read, W.C. Wilbanks, S.J. Schilling, W.L. Bleecker; fourth row, Deane G. Carter, H.R. Rosen, Dwight Isely, Martin Nelson, W.J. Baerg, W.H. Sachs, C.C. Whelan, Harold Barr, R.H. Mason; back row, Ralph Foster, Hubert Hinds, R.H. Austin, H.A. Sandhouse, Lynn Osborn, Edgar Martin, C.W. Rapp, Paul Millar, J.O. Ware, S.R. Stout, O.W. Rowland.

Chapter III

Agriculture in the 1920s

In 1925 with the conflict over the location of the Experiment Station and the University resolved, the legislature moved to appease local interests by committing money for three branch stations. The Rice Branch Station was established in Stuttgart on 160 acres under the supervision of Hartsell Banks. The Fruit and Truck Branch Station was established on 185 acres in Hope under assistant director in charge, G.W. Ware, Sr. And the Cotton Branch Station was started in Marianna with an original purchase of 160 acres and with E.B. Whittaker as the first director. Moreover, the Experiment Station committed itself to work throughout the state on local soil conditions on varieties and cultivation techniques for corn, cotton, rye, wheat, sorghum and soybeans. The Main Experiment Station had expanded its farm to 572 acres and in 1928 moved into a new quarter of a million dollar agriculture building, which is still in use today. In addition, the University constructed the first unit-space dairy barn in America for the Experiment Station's prize-winning dairy herd.

The economic boom in the latter half of the 1920s created an optimistic spirit. The so-called "New Era" of prosperity reflected the belief in science's ability to make society more efficient and productive. By 1928, over 10,000 people had visited either the branch stations or the Main Experiment Station to learn about the advances in scientific agriculture. The state commissioner of agriculture, Earle Page, publicized the activities of the scientists.

During the decade of the 1920s, the ten departments of the Experiment Station were dealing with the major problems in "finding the basis upon which improvements and progress can be made." For example, in agricultural chemistry, Barnett Sure and M.C. Kik were making significant contributions in the relationship between vitamin B complex and health. Sure examined the impact of vitamins on infant and animal mortality, major problems among an Arkansas population that suffered from dietary deficiency. He concluded that vitamin B in rice polish could ameliorate beriberi and anorexia and recommended that nursing mothers take cod liver oil in orange and tomato juice.

In agricultural engineering, which became a department in 1920, Deane G. Carter, an Iowa State graduate and head of the department, developed methods of economically draining wet land and preventing erosion, assessed the efficacy of new farm methods and machinery and worked on unit space barns and simplified self feeders for hogs and poultry. Carter started a series of research publications dealing with the most commonly regarded problems of farmers, including the opportunity for the substitution of tractors for mules, inexpensive housing for people and livestock and preservation of wood fencing. In 1923, the department built a fence around a vineyard to test durability of materials for fenceposts, the value of preservative treatments, the effect of slope and the toxic effect on spe-

In 1925, the Arkansas legislature approved the establishment of three branch research stations. The Rich Branch Station (now called the Rice Research and Extension Center, shown above) near Stuttgart focused primarily on rice and soybeans. The branch station at Hope concentrated on fruits and vegetables, and research at Marianna centered on cotton.

This agriculture building was constructed in 1927 to provide more facilities for research and teaching. It is still the headquarters for the Station director and dean of the College of Agriculture and Home Economics. (Photos courtesy Photograph Collection, Special Collections, University of Arkansas Libraries.)

This dairy building contained offices for the departments of animal and dairy husbandry (now the department of animal and poultry sciences) from 1905 until 1957. It also housed the creamery. (Photo courtesy Photograph Collection, Special Collections, University of Arkansas Libraries.)

cific organisms of certain chemical treatments used as preservatives. H.T. Barr designed an experiment to measure the rotating power of tractors.

In agronomy, under department head Martin Nelson, J.O. Ware, C.K. McClelland, George Janssen, R.P. Bartholomew and L.C. Kapp continued ongoing fertilizer studies and investigated the impact of crop rotation and new seed varieties on yield. Ware studied cotton cultivation practices and crossbreeding to develop better strains of Acala and Rowden cotton. Nelson experimented with fertilizers and appraised the effect of different combinations of fertilizers on cotton crops, trying to develop optimum ratios of nitrogen, phosphorus and potassium adjusted to soil conditions. Janssen and Nelson explored the reasons for the lack of permanency of fall pastures. Janssen also studied the specific difficulties rice farmers had in utilizing fertilizers.

Deane G. Carter, head of the department of agricultural engineering from 1921 to 1941, was the first agricultural engineer to write a Station Bulletin, *Tractors in Arkansas*. He became internationally known for his research in rural housing and kitchen designs. (Photo courtesy Photograph Collection, Special Collections, University of Arkansas Libraries.)

The animal industry department, led by H.E. Dvoracheck, including Edgar Martin, R.M. Smith and M.S. Lippert, determined optimal feeds and "the best practice" of handling food. In entomology, W.J. Baerg, assisted by Dwight Isely and H.H. Schwardt, investigated the cotton boll weevil, red spider, cucumber beetle, housefly and root worm. In plant pathology, V.H. Young, H.R. Rosen and E.C. Tullis worked on cotton wilt, fire blight of apples and pears, black rot of grapes and sweet potatoes, seedling blight and stem rot of rice and downy mildew. In 1925, Congress enacted the Purnell Act to increase funding of the Experiment Station and to increase the scope of the Experiment Station to include rural sociology and home economics by additional annual appropriations of $20,000 with stepped up increments to $60,000 in 1930.

With funding from the Purnell Act, the Experiment Station established the department of rural economics and sociology and hired its head, C.O. Brannen. The present name of the department--agricultural economics and rural sociology--was adopted in 1956.

Brannen began a series of studies on marketing problems of Arkansas farmers. In 1926 he produced his first study on the marketing of Arkansas peaches (Bulletin 207). The peach industry had begun in western Arkansas in the 1890s, and by 1925 Arkansas ranked fourth in the nation in car lot shipments of Elberta peaches to other states. However, the peach industry had experienced hard times in the period between 1910 and 1920 because of crop failure. Limitations on the expansion of peaches included the cost of spraying and the lack of quality control. Another major factor was the division of the nation by the Interstate Commerce Commission (ICC) into different zones of commerce based on the anticipated volume of commerce per mile. Due to Arkansas' underpopulation and lack of major commodities, the ICC allowed the railroad to charge higher rates. The shipment of all Arkansas commodities was charged on a schedule about 40% higher than that for other parts of the United States and about 20% higher than that for the southwest. Consequently, the higher transportation charges became a grievance to Arkansans who claimed the differential left Arkansas a supplier of last resort or a "colony" of the Northeast and Midwest. This economic disincentive resulted in lower exports of peaches to other markets. Arkansas peach producers had to both become more efficient and create a quality niche in the market.

In 1926 Brannen studied the credit system of Arkansas agriculture (Bulletin 208). The establishment of the Federal Land Bank had made a major impact on Arkansas. Available capital resources in Arkansas increased 166% from $102 million in 1914 to $271 million in 1925. In addition, the creation of the Federal Intermediate Credit Bank allowed for creation of Agriculture Credit Corporations. In spite of increases in the availability of credit, interest to farmers in Arkansas was on the average 3% to 5% higher than in the rest of the nation. Brannen attributed this differential to Arkansas' dependence on the price of cotton and felt that diversification was difficult due to "custom, inertia and other forces." Moreover, bankers were "frequently hesitant" to lend on any other crops. Consequently, the average loan per acre in Arkansas was $1,710 compared with $4,595 in Illinois and $3,167 in Missouri. Finally, most Arkansans had funds "scarcely sufficient to meet credit obligations." Brannen also had misgivings that would later be justified that the method of financing highways in Arkansas would increase farm credit risks. Counties issued road improvement bonds passed on appraised values that were subject to fluctuations.

In 1928 with the problems of agricultural surplus depressing prices, B.M. Gile of the department of rural economics and sociology turned his attention to the question of farm credit. No longer self sufficient, the farmers of the 1920s were being increasingly integrated into market economy. Gile emphasized that Arkansas farmers would have to borrow more and more money to put in a crop.

The concept of debt had become outdated. Despite efforts to diversify, Gile reported, 76% of the land in southeast Arkansas was in cotton. He observed that tenants, due to their circumstances, did not fully apply themselves to the land. Conditions were not any better in the piney woods area where farmers of small farmsteads without any financial resources were on the brink of poverty. Tenants fed their "half-domestic hogs" on corn and raised cotton during the summer, supplementing their income by hauling ties and firewood or by working in local sawmills.

Dan Gray, the head of the Experiment Station in Arkansas, focused the Station's resources on the problems of cotton. Entomologists Dwight Isely and W.J. Baerg concentrated on the boll weevil, which had destroyed some 21.84% of the cotton crop in 1921 (Bulletin 190). Arkansas' shorter growing season, more severe winters and lower relative humidity had resulted in only half as much crop damage as was experienced in Georgia and Oklahoma. Although the weevil entered the South in the early 1900s and Arkansas in the early teens, it did not become a major pest until 1921.

Dwight Isely believed that biological control might offer a major impediment to boll weevil proliferation. Since the weevil hibernated in the winter, Isely advocated destruction of all shelters such as dead grass, ditches, terraces, fence rows, blackberry hedges and brush along fields as well as the cleaning of all gins and destruction of cotton stalks before spring. Another approach was the development of early-blooming varieties of cotton. These varieties would also protect against the cotton leaf worm, the army worm, the cotton bollworm and the red spider. By 1922, Arkansans were dusting plants with calcium arsenate and stripping squares from the cotton plants. Concluding that tenants were applying the poisons too soon or too late, Isely crisscrossed the state preaching his ideas on control and emphasizing the proper timing of applications. However, Isely realized that his ideas would not be easily adopted. Many tenants refused to touch arsenate compounds, rain washed the poisons out, and the price of calcium arsenate went up dramatically in the summer. Also, the poison had the unanticipated consequences of favoring the development of cotton aphids.

The prosperity of the 1920s led to a modest revival in cotton prices. In 1926, Isely's ideas regarding biological control of the boll weevil were to be practically tested on two plantations near Pine Bluff. Despite problems with labor management and from unfavorable weather, boll weevil infestation was repressed during the grow-

The Home Economics building, similar in design to the Agriculture building, was built in 1939. The Purnell Law passed by Congress in 1925 helped support research in home economics and rural sociology. In the late 1920s and 1930s, home economists studied vitamins and health services in rural areas. Later studies involved clothing, home improvement and rural housing.

ing season. However, Arkansas cotton was hit with three new pests: the cotton hopper, the tarnished plant bug and the cotton leaf bug, which had originated in Texas. Isely recommended the eradication of weeds and evening primrose as hosts.

Isely advised that the southern corn rootworm be dealt with by destroying wild grasses a month before planting to avoid infestation (Bulletin 232). Isely warned against planting corn if heavy precipitation occurred prior to June. In dealing with boll weevils, Isely, after careful study of the life cycle of the pest, made the observation that the reproduction of weevils was dependent upon squares or small bolls being available for food. Consequently, an approach to insect control could be seen in insuring that the weevil had no food or shelter to survive the winter. Isely's associate, W.J. Baerg, investigated the life history of tarantulas, shade tree insects and the rough headed corn stalk beetle.

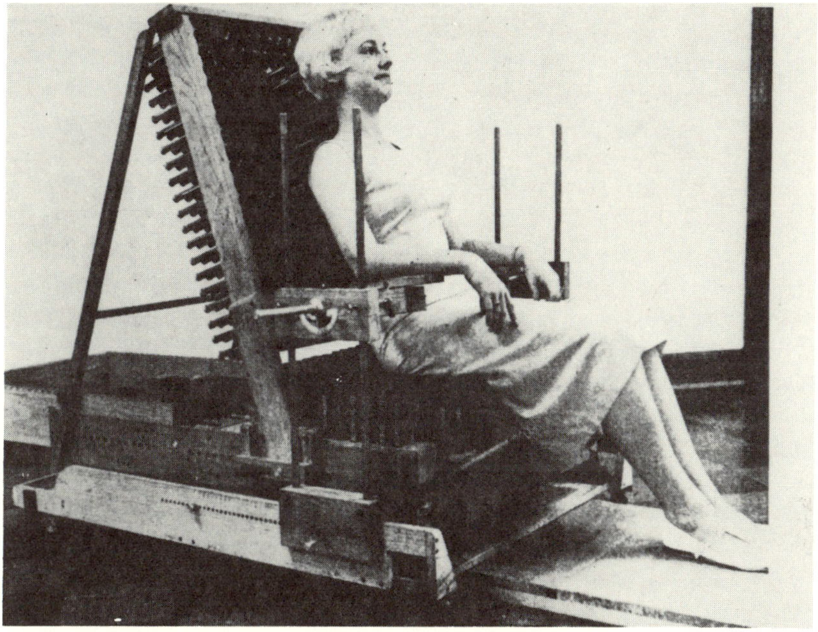

The photograph above was reproduced from Station Bulletin 616, *Basic Design Measurements for Sitting* by Clara R. Ridder, published in 1957. More than 30 years later, it is still one of the most requested Station bulletins.

Development of improved cotton varieties remained a major priority for breeders. A goal of breeders was to develop a cotton with sufficient earliness to avoid severe boll weevil infestations, in particular in the shorter season of the northern Delta where a large percentage of the Arkansas cotton crop was produced. About 1925, J.O. Ware developed a new strain of Rowden cotton, designated as 'Arkansas Rowden 40', followed a few years later by 'Rowden 2088' and 'Rowden 5056'. Other strains of Rowden cotton later developed at the Station included 'Rowden 41A', 'Rowden 41B', 'Rowden 42A', 'Arkot 1', 'Rowden B5', 'Rowden B28' and 'Rowden 60A'. Later Ware discovered that red plant color, okra-shaped leaves and green lint color in cotton were independently inherited characteristics. Ware concentrated his efforts on the problem of breeding cotton that was resistant to wilt.

As the stock market crash of 1929 approached, Arkansas' economy was already under stress. For cotton growers, the decline had begun in 1920. Middling cotton sold in the Little Rock market in

Cotton research intensified in the 1920s and 1930s as farmers and researchers battled the boll weevil and other insect pests of cotton. Agronomist J.O. Ware, shown above, developed improved cotton varieties, and entomologist Dwight Isely advocated biological control practices, proper timing of chemical applications and the eradication of weeds.

June 1920 at 40 cents/lb, dropped to 13 cents by December 1920 and 8 by early 1921. In addition, some planters had carried over the 1919 crop into 1920 in anticipation of higher prices. As a consequence, many cotton growers were in deep trouble, having borrowed heavily to make a crop.

The result of financial distress was the formation in 1922 of the Arkansas Cotton Growers' Cooperative Association. The idea was to bypass the factors that were believed to downgrade cotton and to sell good market cotton to the mills in an "orderly" fashion. The first year of the association was unfortunate. The association had 200,000 bales under contract, but only 70,000 bales were delivered. B.M. Gile undertook to try to understand how the realities of the cotton economy thwarted the government's efforts to raise cotton prices by a study of the *Status of Cooperative Cotton Marketing in Arkansas* (Bulletin 245). Gile's research indicated that by 1929, membership in the cotton cooperatives had declined one-third from its 1926 peak when the association had claimed 18,600 members. Gile pointed out the weakness of cooperatives in enlisting members and encouraging their participation, including the fact that the association insisted on five-year marketing contracts, regardless of satisfactory performance by the cooperative. Gile referred to the "spirit of individualism" that rendered cooperative action difficult and the thrill the farmer derived from selling his own cotton.

In 1926, a record-breaking crop was picked, and cotton prices declined. The association was forced to dump its cotton on the market for 8 to 9 cents/lb as farmers demanded immediate payment in order to liquidate old debts. The association reduced marketing agreements to two years, but this change did not affect antagonistic attitudes towards the association. In 1929, Congress enacted the Agricultural Marketing Act to assist the cooperatives financially. A Farm Board was formed to advance money to cooperatives such as the Mid-South Marketing Cooperative. The Farm Board worked with the Federal Intermediate Credit Bank to advance to members loans on cotton. The availability of credit led to an unanticipated increase of cotton in storage. As a result of the unforeseen rapid collapse of domestic demand combined with foreign trade restrictions, the Farm Board, with its large unsold store of cotton, became part of the problem rather than the solution.

In other areas of research, scientists promoted the nascent livestock industry and continued long-standing investigations into animal nutrition. In the 1920s, agricultural experts emphasized raising dairy cows and hogs as an important part of the "live-at-home" philosophy. Consequently, Experiment Station scientists worked on silage for dairy cows. The Experiment Station told farmers that corn mixed with soybeans provided the best silage for swine and that use of peanut meal, soybeans and cottonseed meal could provide satisfactory rations. Most hay bought by Arkansas farmers was grown in Oklahoma and Texas and consisted of native grasses and weeds with deficient nutritional qualities.

H.E. Dvorachek continued his experiments testing the value of legume hay for dairy heifers (Bulletin 196). Dvorachek investigated the question of whether legume hay was more valuable than grass hay due to its higher protein, mineral and vitamin content. After tests on six cattle, Dvoracheck concluded that alfalfa showed no advantage. In addition, the animal husbandry department undertook major cooperative research projects with the U.S Department of Agriculture and the state Agricultural and Mechanical College at Jonesboro to test the efficacy of feed lots.

Another area that the Agricultural Experiment Station started to explore was improvement in the poultry industry, an endeavor of substantial benefit to Arkansas. R.M. Smith's first tests involved the use of feed supplement in a study to examine hatchability and fertility. Smith concluded that chickens could be enclosed in a house and fed a basic ration of sprouted oats or alfalfa leaf meal, which would increase percent fertility and hatchability of eggs. However, cod liver oil supplementation was not necessary unless rice byproducts were used as a substitute for corn, with the accompanying loss of vitamin A.

Improvement in cultivation techniques was important to increasing yields. Martin Nelson, assistant director of the Station and head of the department of agronomy, continued his experiments with corn cultivation, which began in 1909, at Fayetteville, Imboden, Danville and Mena. Given the vagaries of Arkansas weather and the great varieties of soil type, late-maturing corn seemed to do best, especially if the rows were not planted too close together. Increasing knowledge of soil types allowed the Experiment Station to guide farmers based upon experiments and allowing for the erratic rainfalls in Arkansas. Nelson emphasized the importance of insuring available nitrogen and winter cover crops to avoid erosion and restore nitrogen to the soil.

Station scientists had an ongoing project to promote the use of Irish potatoes as a dietary supplement. In 1918, the Experiment Station had leased land with a portion reserved for potato experiments. The experiments demonstrated that phosphoric acid, nitrogen and potash at a ratio of 8-4-4 gave the best results when applied at planting time. Nitrogen furnished by nitrate of soda gave better results than that furnished by cottonseed meal. However, in cash-poor Arkansas, few farmers could afford fertilizer. To assist poorer farmers, Martin Nelson started to determine standardized fertilizer mixtures that could be home-mixed.

Another research area initiated as a result of funding from the Purnell Act in Arkansas was farm home planning. Only one third of the tumble down shacks of tenants in the state were painted, and more than half did not have sanitary facilities. The only modern labor-saving device possessed by most families was a sewing machine. Only 2% of the farmsteads were electrified. Home economists were hired by the Experiment Station to work with other agricultural scientists on upgrading rural life.

The major objective of the Experiment Station was to help farmers to live at home and be self-sufficient. The Experiment Station believed that acreage yield must be held at a high figure to reduce the amount of feed purchased. In order to achieve this objective, extensive tests were conducted to test new varieties and different approaches toward cultivation techniques.

In horticulture, J.R. Cooper and his associate C.B. Wiggans continued the cross pollination studies of Ben Davis and Delicious apples. The horticulturists examined fertilizer and studied measurement of trunk growth, pruning and performance of fruit buds. They concluded that there was a definite positive correlation between tree vigor and crop production. In 1928, studies were commenced in Nashville, Arkansas, on fertilizing peach orchards. Wiggans started studies of Concord grapes already being cultivated in the Arkansas Valley and at Tontitown, a small Italian community north of Fayetteville. New varieties of grapes were grafted, and tests were conducted on pruning procedures.

J.R. Cooper and V. M. Watts conducted fertilizer experiments with sweet and Irish potatoes with results indicating that phosphorus and manure caused a major difference in yields. They recommended applying the fertilizer several days before planting with the optimal ratio of 10-5-5.

First Soybean Variety Developed by the Arkansas Agricultural Experiment Station:

ARKSOY, 1928

The department of horticulture also worked on improved pollination of apples. Work on fertilizers in apples and peaches indicated that nitrogen should be applied to trees at blooming time. Moreover, cover crops should be utilized to supply "green manure."

V.H. Young in plant pathology concentrated his efforts on cotton wilt fungus, the most devastating disease of cotton, finding that susceptibility could be reduced by the effect of low soil and air temperature. Consequently, breeding experiments were conducted by agronomy and plant pathology professors Young, Ware and Janssen to try to breed a wilt-resistant variety. H.R. Rosen also believed that the use of manure contributed to wilt. Another major project of Rosen's was his attempt to disprove the belief that pear trees were primarily responsible for carrying the fire blight disease that ravaged the apple industry. In his search for the transmission mechanism, Rosen surmised that honeybees might be involved in the spread of the disease. Rosen also worked on tip and margin burn of Irish potato leaves and concluded that the diseases were caused by leaf hoppers. Rosen recommended the Irish Cobbler variety rather than Bliss Triumph because of its disease resistance.

Chapter IV

Agriculture and the Great Depression

On the verge of the Great Depression, the Experiment Station continued the broad lines of inquiry that had commenced in the mid 1920s. The department of agricultural chemistry concentrated on problems of vitamin B deficiency, which had been identified as a major cause of pellagra and beriberi. Margaret Elizabeth Smith and Barnett Sure explored the importance of vitamin B for lowering infant mortality. The scientists continued work on a concentrated, potent form of vitamin extracted from rice polish. Deane Carter in agricultural engineering worked on planning inexpensive housing for tenants and continued studies to test the effect of toxic chemicals on wood posts. In addition, agricultural engineers studied ways of utilizing high efficiency electric pumps for irrigation of rice fields.

In agronomy, Martin Nelson and J.O. Ware continued studies on the efficiency of combinations and application of fertilizers in cotton, fruits and vegetables in order to determine what could be used most economically and to determine the point of diminishing returns. Nelson suggested that cottonseed meal as a fertilizer was effective in preventing caking in mixed fertilizers. Nelson's research indicated the importance of crop rotation involving corn, oats, wheat and clover. Tests continued on varieties of soybeans and alfalfa and on possible fertilizers for rice. R.P. Bartholomew and George Janssen conducted experiments on increments of potassium fertilizer utilizing a range of crops including tomatoes, corn, cowpeas, soybeans, oats, wheat, cotton and grasses to determine the necessary levels of potassium for healthy plant development.

In entomology, Dwight Isely continued to call for winter clean up to cut down on boll weevil infestation. Moreover, Isely cautioned against overuse of fertilizers, which increased the vegetative growth of the cotton, consequently encouraging the weevil. The Experiment Station in Lonoke started to research the sawtoothed grain beetle, which infested polished rice byproducts.

In horticulture, scientists explored the techniques of cross pollination and the effect of fertilizers and pruning on fruit trees. Other crops included in fertilizer tests were grapes, strawberries, potatoes, tomatoes and cantaloupes. In addition, guidelines were developed on fertilization of cantaloupes, watermelons, radishes, asparagus and peppers. In 1929, the Experiment Station started greenhouse experiments with fertilizer and artificial light. New experiments in forestry concentrated tests on root growth, seed germination and wood studies on a 26-acre tract set aside on the Fruit and Truck Branch Station in Hope.

In the field of disease control, a major concern was in controlling blossom blight by applications of germicides. E.C. Tullis commenced research on stem rot in rice and began studies of biological control for rice diseases, including blight and smut. The Experiment Station suggested draining as a partial control of smut.

In November 1930, the Arkansas banking system collapsed as the American Exchange and Trust in Little Rock, with 72 branches

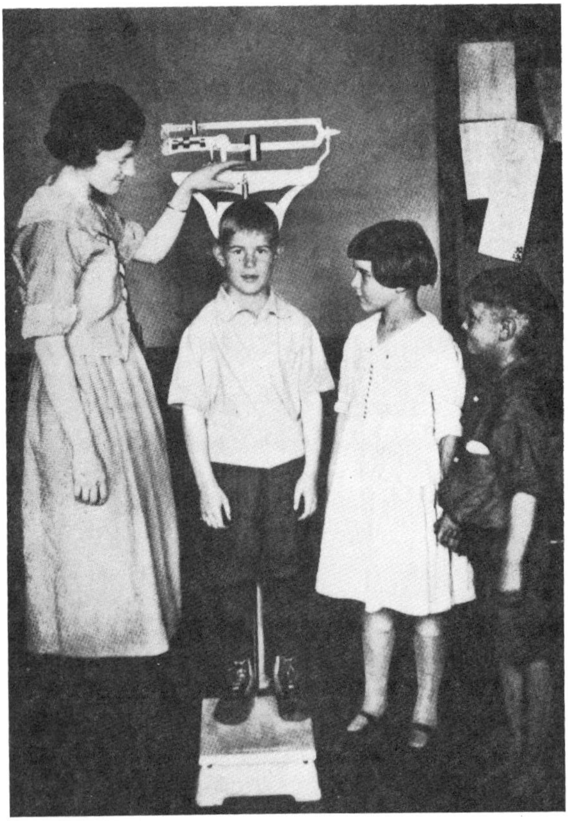

During the 1920s, vitamin deficiency was a serious problem among many rural residents in Arkansas. This photograph was reproduced from Station Bulletin 184, *Vitamins, Health and the Daily Diet*. The authors, J. W. Read and Stella Palmer, discussed the importance of vitamins and minerals in the diet and suggested various home and school menus.

throughout the state, closed its doors. As a consequence, most banks in the state closed. In the face of the mounting bank crisis, B.S. Gile continued his studies of rural credit and observed that farmers and tenants had increased their debt burden 150% during the 1920s (Bulletin 259). However, the interest rate paid by each group differed considerably with farmers paying 13.4% and tenants 17.8%. Rural merchants by-passed the Arkansas 10% usury law by adding a flat rate of 8 to 10% to the cash price of goods sold on time. In spite of a recovery in cotton prices, 51% of farmers could not pay off their accounts. Farmers fell into tenancy by their inability to redeem the lien on their cotton on "settlement."

The insufficiency of credit stifled all aspects of farming. Rural economist Carlos Campbell analyzed the popular, small-scale business of tomato growing in western Arkansas and found that, as a result of growers' lack of economies of scale and lack of centralized buying and selling, canners were overladen with debt and had to dump a large part of their harvest in the fall, depressing the price (Bulletin 240).

The only crop that maintained its price was the highly capitalized rice cultivation on the Grand Prairie. O.J. Hall, analyzing rice industry returns, pointed out that the farmers' net receipts averaged $1,995 in contrast to $254 for cotton farmers. The rice areas of the state stood in economic contrast to the areas of the prevailing cotton culture although in close geographical justaposition. Rural sociologists and economists studying the rice belt of Arkansas saw Arkansas and Prairie counties as islands of prosperity in a sea of poverty. DeWitt and Stuttgart reflected the higher income of operations by the high rate of automobile ownership. Moreover, most families were active in cooperative associations, home demonstration clubs and churches. Life in those counties seemed to fulfill the hopes of agricultural scientists.

In the summer of 1930, Arkansas agriculture suffered a devastating blow due to a terrible drought combined with a decline in the price of cotton. The drought ruined corn and legume crop experiments and destroyed pasture crops. Agronomist R.P. Bartholomew cautioned farmers about the importance of killing weeds to preserve soil moisture and nitrogen. He wrote that winter cover crops could prevent an average loss of 50 lb of nitrogen/acre. However, in a farming system in which one-third of the sharecroppers moved every year, farmers found it difficult to follow the advice of agronomists.

C.O. Brannen, the assistant director of the Experiment Station and a rural economist, warned farmers of the need to maximize their

During the Depression, Barnett Sure and M.C. Kik traced the role of rice in combatting vitamin deficiency and helped to reduce infant mortality. This 1950 photograph shows faculty members in the department of agricultural chemistry. First row (l to r) Joy Dowell, Barnett S. Sure and Mary Crudup. Second row Leslie Easterling and M.C. Kik.

limited resources by utilizing the information provided by the Station as a basis for the necessary economic recovery of the state. Without sound agricultural practices, Arkansas farmers could not compete in the regional or national markets.

The Experiment Station work continued along the lines that had been developed in the late 1920s. Barnett Sure and his associates in agricultural chemistry worked on problems of vitamin deficiency. Due to the inadequacy of the diet of the poor, vitamin deficiencies, especially pellagra, were pandemic.

The agricultural engineering department conducted research over a range of projects. Deane Carter worked with cost analysis of 2- and 4-row cultivator equipment to try to lower cultivation costs. The Experiment Station continued tests on the best method of preserving posts and analyzed rice cultivation to determine the efficiencies of electric versus oil-drive engines.

One of the major research programs was Martin Nelson's long-range determination of fertilizer levels. Nelson emphasized that 100 lb of fertilizers could increase yields from 100 to 150%. Twenty pounds per acre of nitrogen derived from cottonseed meal seemed to be optimal in stimulating earliness of fruit and maturity. On the poorest of soil, super phosphate seemed to be the most efficient in restoring productivity. Nelson also grappled with the problem of fertilizing rice, which did not respond to fertilizer as favorably as other crops. The greatest increase in rice yield per acre seemed to be from better varieties such as 'Calora Japan' and 'Supreme Blue Rose'. In controlling weed production, Nelson recommended alternating rice with a non-irrigated crop, such as soybeans.

J.O. Ware continued his testing of the most desirable varieties of cotton. Ware disparaged 'Half and Half', a favorite of many farmers, as not "measuring up to expectations." The Experiment Station urged the use of 'Arkansas Rowden 40'. The Cotton Branch Experiment Station in Marianna was the site for the testing of cotton free from disease that had qualities including size of boll, desirable staple length, ease of picking and storm resistance. It took five years of selection to choose advanced strains. After selection, 100-lb bags of seed were sold to farmers through county agents.

The use of agricultural byproducts was a constant concern of Station scientists. Edgar Martin conducted soft pork investigations in animal industry. Rather disappointingly, he concluded that rice polish and rice bran could be fed only to pigs with an initial weight of over 75 lb. Smaller pigs with initial weights of 40 to 50 lb did not flourish on rice byproduct rations. H.E. Dvoracheck concluded that cod liver oil did not improve profitability since it was relatively expensive. In addition, the Experiment Station warned that the use of rice byproducts, brewers rice, rice bran and rice polish with or without cod liver oil supplements was not adequate. Dvoracheck also tested rice polish and cottonseed meal as feed for dairy heifers. Similar nutrition experiments using rice byproducts proved more successful with chickens.

In cattle nutrition experiments, prairie hay was compared to alfalfa to examine comparative nutritional qualities. Dvorachek concluded that prairie hay with a balanced supplement could produce results similar to alfalfa hay but with the provision that during lactation, minerals and vitamins were necessary.

The department of rural economics and sociology reported that, in all areas of agriculture, Arkansas farmers had experienced a decline in cash flow. This income reduction, especially in cotton, was due to the 40% fall in cotton prices from 1924 levels. Reflecting greater returns to scale of production, only large producers in rice showed a profit. By 1931, the agricultural and financial depression in Arkansas reduced University funding. Consequently, the Experiment Station could not undertake new scientific initiatives. With the crop

Dusting cotton with a mixture of powdered lead arsenate and hydrated lime was a common practice to control insects.

failures and subsequent displacement of sharecroppers, governmental reports and newspapers spoke of deprivation and starvation among both animals and people. Agricultural engineer Deane Carter predicted that the substitution of tractors for mules would adversely impact sharecroppers. Furthermore, the drought of 1930 had led to a major decline in the groundwater level from which rice irrigation was drawn. In some areas the water table fell from 20 to 35 feet.

The Experiment Station embarked upon an exhaustive examination of the problems of fertilization of rice. Rice scientists demonstrated the relation of exchangeable iron and manganese in submerged soil to plant growth. Martin Nelson had made considerable progress in developing effective fertilizers for rice. Ammonium sulphate and sulphate of potash seemed to provide promising results. Working with the U.S. Department of Agriculture, the Stuttgart Branch Station tested varieties of rice to explore earliness, size of grain, time of blooming and appearance.

George Janssen started experiments to study the absorption and utilization of calcium nitrate and ammonium sulphate by the rice plant. Janssen and L.C. Kapp concluded that the three necessary elements for rice growth were nitrogen, iron and potassium. To

COTTON VARIETIES AND STRAINS DEVELOPED BY THE ARKANSAS AGRICULTURAL EXPERIMENT STATION BEFORE 1940

Arkansas 17	Acala 31
Arkansas Rowden 40	Acala 34
Arkansas Rowden 2088	Acala 37
Arkansas Rowden 5056	Acala 892
Arkansas Rowden 41A	Arkot 1
Arkansas Rowden 41B	Arkot 2-1
Arkansas Rowden 42A	"Deltapine" A8
Arkansas Rowden B5	"Deltapine" A12
Arkansas Rowden B28	"Deltapine" A40
Arkansas Rowden 60A	

sustain soil nutrition, a two-year rotation alternating rice with soybeans appeared to be effective.

In the area of cotton cultivation, experiments continued on effects of seed variety, spacing of plants and soil type on fertilizers. By 1930, many progressive farmers were utilizing the three seed varieties developed for cotton in conjunction with the Experiment Station: 'Arkansas Rowden 40', a general purpose cotton; 'Arkansas Acala 37', an early strain bred for the northern portion of the Arkansas cotton-growing sections; and 'Arkansas 17', for the Delta.

The national economic crisis worsened in 1931, paralyzing state financing. Confronting a decline in tax revenues, Arkansas defaulted on its highway bonds and suspended cash salary payments to teachers. Emphasis was placed on making do with scant financial resources.

Scientists at the Experiment Station carefully researched the most cost-efficient methods to assist agriculture. Entomologists sought control of the garden webworm that attacked alfalfa and cotton. Dwight Isely pursued his ideas of biological control of the strawberry weevil, which could be managed by cleaning and burning during the winter, especially the borders of strawberry plots in woodlands where weevils hibernated. W.J. Baerg recommended dusting with a mixture of one part of powdered lead arsenate to three or four parts of hydrated lime.

The horticulture and forestry department continued fertilizer tests on apples, grapes, peaches, potatoes and cantaloupes. The results were below what were expected, but weather conditions interfered in testing. New varieties of peaches, nectarines, black walnuts, tomatoes, pecans, English walnuts and filberts were also tested.

Plant pathology concentrated on stem rot and sterility in rice as well as cotton wilt. Tests on rice sterility, or straighthead, concluded that aerated soils containing relatively large amounts of organic matter tended to produce sterility. In combating cotton wilt, research indicated that liberal applications of fertilizer containing potash could ameliorate an extremely severe type of "rust," or potash hunger. In addition, analyses indicated that potassium eliminated the cotton rust symptoms.

In dealing with fire blight in apples, plant pathologists faced a crisis situation; losses statewide averaged 30%. In Ozark, Arkansas, a whole orchard of 400 21-year-old Jonathan apple trees was entirely infected. Pathologist H.R. Rosen continued work on bees as the suspected vectors for the spread of the blight.

The department of rural economics and sociology chronicled the enormity of the agricultural crisis. Of the 22 agricultural credit

In the 1930s plant pathologists concentrated on controlling stem rot in rice, cotton wilt and many other disease affecting both agronomic and fruit crops. This photograph from Station Bulletin 330 *Studies of Oversummering of the Fire-Blight Pathogen* by H. R. Rosen, shows a Jonathan blossom affected by fire blight. In one year, fire blight destroyed 30% of the state's apple crop.

corporations in Arkansas, 15 had failed. Only 29.7% of the farmers in Arkansas showed any profit in 1930. With no relief from property taxes, tax delinquency rose dramatically. As a result, Arkansas tottered on the brink of default of its state debt due to the decline in revenues and the overcommitment to the building of farm-to-market roads.

Conditions in the state continued to deteriorate in 1932. The public school system and state government operated on state warrants of promised payments at redemption. Research aims were limited to already-existing lines of inquiry, and the research on rented land in central Arkansas was suspended. In Fayetteville, the Station devoted 32 acres to corn, 28 to legumes, 21 to fertilizers, 24 to small grains, 2 to sorghum, 2 to cotton breeding and 24 to weed and pasture. Horticultural tests occupied 85 acres of which 45 acres were in fruit trees, 10 in grapes, 15 in truck crops and 15 in forestry and nut planting. About 2 acres were given to entomological and pathological research relating to apples.

The continued interest in agronomy was on lower-cost fertilizer options. Nelson renewed emphasis on cottonseed meal as a fertilizer and suggested that it could be used to the extent of 10 to 30% of the total nitrogen in fertilizer.

The agricultural chemistry department continued to focus on vitamin A and B deficiencies, common problems in Arkansas, especially in the wake of the drought. Barnett Sure and H.S. Thatcher were particularly concerned with finding an easy test to detect vitamin deficiency.

Agricultural engineering dealt with a multitude of problems such as mechanization and rural housing. Deane Carter predicted that cotton production could eventually be easily mechanized for all tasks except chopping and picking. Animal industry continued experiments on hog feeding, and tests with soybeans and corn appeared promising. Similar tests were also done on laying hens, especially to see if rice byproducts could be used as feed.

W.J. Baerg in entomology continued work in biological control. In 1931 Baerg cautioned farmers to avoid damage caused by the rough headed corn stalk beetle by planting corn in land that had not been in sod and ensuring proper drainage. Similarly, the sawtoothed grain beetle could be controlled by cleaning polished rice. In dealing with horse flies, entomologist H.A. Schwardt recommended cleaning ponds. Dwight Isely studied the life cycle of the rice water weevil and refined the technique of timed draining of the rice fields, recommending that lowering water levels approximately three to five weeks after planting would destroy the larvae during the most critical period in the insect's life history. Finally, in dealing with a perennial pest, the boll weevil, Isely concluded that humidity played a major role in the success of boll weevil infestations. He observed that the devastation created by the drought of 1930 had practically wiped out the pest. However, by 1932, with the return of the normal rain cycle, the boll weevil was making a comeback.

V.M. Watts, George Janssen and R.P. Bartholomew continued their long-term analysis of pruning and fertilizing of apples. The experiments had started in 1915 on fertilization of apples, particularly Winesaps, Ben Davis, Jonathan, Delicious and Golden Delicious. Cooper and Wiggans emphasized the importance of continual use of fertilizer, in spite of high prices, in order to maintain healthy orchards. Also, guidelines were developed for optimal use of cultivation for asparagus, snapbeans, cantaloupes, cucumber, sweet potatoes, Irish potatoes, tomatoes and watermelons.

Rural sociologist B.M. Gile published startling figures on the scope of the impact of the agricultural depression. The amount of seasonal credit in Arkansas had declined from $71 million to $45 million in 1931. Farmers were unable to pay taxes in many counties. Agricultural prices, especially for cotton, fell to the level of 1921 while many farmers in eastern Arkansas did not even make a crop. Livestock had been particularly hard hit. C.O. Brannen viewed the crisis as the most critical ever to confront farmers and predicted that, without modernization and diversification, many operations would go under.

In a major Station report, Brannen admonished farmers that efficiency could not be neglected at the expense of a low standard of living or failure. To justify continued state support of the Experiment Station, Brannen pointed out that state cotton growers had earned $1.6 million as a result of improved varieties. Despite the low prices for cotton, farmers were reminded that yields would diminish without adequate fertilizer. Dwight Isely warned farmers that the introduction of cotton pests from other parts of the world combined with continuous cropping of cotton and no soil improvement could result in rapid spread of cotton pests throughout the state.

To compound agricultural problems, the rapid decline in farm income was not accompanied by a concomitant reduction in property tax. As a consequence, farmers confronting a tax bill sometimes

C.O. Brannen, a specialist on taxation and land tenure, issued a comprehensive report in 1933 emphasizing the major contributions of the Experiment Station to Arkansas agriculture and warning of dire consequences of the failure to adapt recommendations. Brannen was appointed head of the department of agricultural economics and rural sociology in 1927 and served as assistant director of Station research from 1928 to 1944. He then became director of the newly formed Bureau of Research at the University of Arkansas until he resigned in 1948.

larger than net revenues were forced into tax delinquency. By 1932, the state had a major problem in the disposition of defaulted property. Experiment Station researcher Orville J. Hall worked with state officials and the U.S. Department of Agriculture to study the state's tax delinquency program and recommend the disposition of land that had reverted to the state.

In 1933, the Roosevelt Administration enacted the Agricultural Adjustment Act to pay farmers not to put into cultivation a portion of their wheat, cotton, field corn, rice and tobacco. In addition, the Commodity Credit Corporation was established to allow farmers to borrow money on crops and put corn or cotton up as collateral. Farmers responded by cutting corn by 50% and cotton by 75%. In the cotton South, the program had the unanticipated effect of forcing the eviction of tenants whose labor was now superfluous.

The agricultural relief program in Arkansas became involved in the larger, more difficult question of the unstable agricultural base in the Arkansas uplands, which resulted in long-term poverty of isolated farmsteads. Throughout the 1920s, families in the Ozark and Ouachita Mountains had been emigrating. With the clearing of the forest cover of the Mississippi Delta and its tributaries, the Mississippi, White, Red and St. Francis rivers, wealthy northerners came into the region to establish cotton plantations. The collapse of the price of cotton led to the end of paternalistic efforts to shelter tenants during the winter months. Landowners advised tenants to seek aid from the federal government. As a result, the federal government shouldered the problem of rising agricultural prices as well as the question of tenant assistance.

C.O. Brannen and other rural economists and sociologists on the Experiment Station staff worked with the U.S. Department of Agriculture to assess the full dimension of agricultural progress. In 1934, Congress put compulsory restrictions on cotton acreage. Between 1933 and 1935, the Agricultural Adjustment Administration reduced the output of cotton by over 10 million bales.

The disability of Arkansas agriculture due to low income and unhappiness with the functioning of the Agricultural Adjustment Administration resulted in the creation in 1934 of the Southern Tenant Farmers Union (STFU). The STFU quickly attracted national attention due to the efforts of the Socialist Party of America. In response to criticism directed at President Roosevelt for his inaction over increased farm protests and violence in 1935, the Research Division of the Work Progress Administration and the Bureau of Agricultural Economics working with the Experiment Station conducted a series of studies on conditions in eastern Arkansas. The work, conducted by K.H. Hunter and H.W. Blalock, received national scrutiny and became the basis for the creation of the Farm Security Administration in 1937. Blalock and Hunter found that tenants were often charged 25% or more interest at plantation stores. After all debts were paid back, tenants were left with as little as $80 a year. In spite of governmental policy, the purchasing power of agricultural Arkansas was continuing to decline relative to the rest of the United States. Only by switching away from cotton could farmers hope to improve their incomes. The researchers confirmed charges of the ineffectiveness of existing federal policies to aid in the recovery of rural America.

The State Rehabilitation Office employed W.T. Wilson and W.H. Metzler to analyze the functioning of the relief administration in Arkansas (Bulletin 348). The two social scientists investigated complaints of discrimination against black tenants. In the Delta, 97% of the black applicants were turned away from receiving relief in order to keep them as farm labor. Farmers were abandoning their farmsteads to go to town or to other states in search of assistance or jobs. The prostrate rural economy reflected itself in the weakness of social support. Rural school systems were barely open, and banks had virtually ceased to exist. Federal assistance failed to resurrect the agricultural sector due to continued low commodity prices. Many tenants were stigmatized as "dishonest, unreliable and non-cooperative" and did not receive consideration for federal help.

When the Experiment Station's state appropriation was almost eliminated, only the Hatch, Adams and Purnell Act funds kept the Station functioning. The staff continued federal projects, but new research directions were impossible to finance. Nevertheless, the exigencies of new federal programs meant new responsibilities. The Dust Bowl years aggravated existing problems.

Federal officials requested that the Experiment Station help farmers find alternative feed for livestock. Experiments were conducted with sudangrass, sweet sorghum, soybeans and cowpeas, and the possibility of utilizing rice or rice byproducts was examined.

The Experiment Station's expertise was also utilized by the federal government's efforts at farm relief. In 1934, Dan Gray went to work with the National Planning Board, and C.O. Brannen assisted the Natural Resources Planning Board. The Agricultural Experiment Station was asked to work with farmers impacted by the dust bowl, farmers whose summer grazing was already exhausted. James Jardine of the Office of Experiment Stations requested that Brannen tell farmers how to feed their cattle in the face of major

The photograph above shows a small, portable mill operated by one family in the 1930s. The operators of portable mills purchased timber in blocks and moved their mill to the timber. In 1932 the Station suggested that the creation of a State Forestry Commission could make Arkansas' timber resources more productive through education and fire protection, prevention of overgrazing and the introduction of more valuable commercial species of pine rather than scrub oak. (Photo courtesy G. Laurin Wheeler.)

shortages of forage, which had been exacerbated by the transfer of cattle from Oklahoma and Texas.

In 1935, Congress enacted the Bankhead-Jones Act to serve as a stop gap effort to continue acreage restrictions. As part of the measure, new additional funding was included for the Experiment Stations based upon a complex formula. Congress would allot an additional $1 million a year starting in 1936 until a total of $5 million was reached in 1940. Sixty percent of the money would be distributed on the basis of each state's proportion of the rural population of the United States. States were expected to match the federal allocation on a dollar-for-dollar basis. The object of the funding would be basic research into new varieties as well as prevention of the denigration of soil fertility.

The Arkansas Experiment Station utilized its additional money in 1937 to establish a fourth branch station in Batesville, encompassing 3,042 acres to study livestock, forestry and land use problems. By 1939 the Civilian Conservation Corps with the aid of the Soil Conservation Service (SCS) had restored the land. In 1938 the Delta Branch Station was established on 100 acres near Clarkedale.

The Soil Conservation Service had been created in 1935 to deal with the devastation of the dust storms of the Great Plains. The new service commenced the study of grasses adaptable to Arkansas

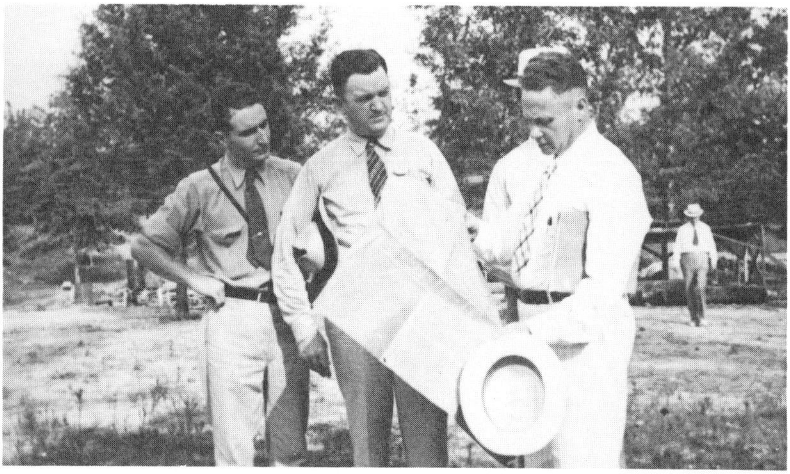

Gov. Carl E. Bailey, shown in the center of the above photograph, visited the Livestock and Forestry Branch Station in Batesville in the fall of 1937. It was under his administration that the branch station was established.

soil conditions. At the Station, R.P. Bartholomew worked on a detailed soil map to encourage an optimal land utilization pattern, and Martin Nelson's studies of soil fertility and erosion control were applied by the SCS.

In 1936, drought struck Arkansas again. In response, Nelson began studies of the feasibility of crop irrigation. The reduction in available acreage stimulated farmers' interest in higher-yielding varieties of cotton. L.M. Humphrey intensified the Experiment Station's efforts to increase yields in cotton.

The Depression and drought years had nearly wiped out the apple industry in northwest Arkansas. Strapped for funds, farmers had done little to protect their orchards. Despite the advice of Dwight Isely in the department of entomology on the control of the codling moth and other pests, farmers despaired of ever recapturing their lost markets. They began to search for alternatives such as poultry, cattle and vegetables. Deane Carter and Robert Smith worked on an improved design for a chicken house. W.R. Horlacher began experiments to crossbreed chickens in order to breed out "bare backs," chickens whose appearance was commercially unattractive. With lack of rain stimulating the proliferation of grasshoppers, entomologist Dwight Isely studied methods for their control.

After years of depression and drought, Station researchers searched for alternative enterprises, such as poultry. Poultry researcher Robert Smith (right) started experiments in poultry nutrition, agricultural engineer Deane Carter improved the design of chicken houses, and W. R. Horlacher began experiments to crossbreed chickens. (Photo courtesy Washington County Historical Society.)

The home economics department, under the leadership of Margaret Smith, initiated a comprehensive survey of the nutritional status of students. Isabella Wilson and William H. Metzler surveyed the abominable health conditions of the rural population and documented poignant accounts of the poor who depended upon herbal and homemade remedies due to their inability to pay physicians (Bulletin 353). Inadequate medical care contributed to high infant mortality. Rural Arkansans often suffered from pneumonia, influenza, malaria, stomach and heart trouble. Fifteen percent of the black population suffered from malnutrition and 14% from eye trouble. These long-term studies stimulated interest in governmental assistance for rural health care.

Research scientists geared their efforts to assisting in the creation of a new Arkansas agriculture more suited to market demands. The thrust of the New Deal agricultural programs called for not only acreage restrictions but also use of alternative crops. In 1928, 'Arksoy', a soybean adapted for Arkansas conditions on former cotton land, had been released by C.K. McClelland. Stock breeding was encouraged. W.R. Horlacher started sheep breeding with Hampshire ewes. Improvement in livestock practices was sought with the development of sorghum for silage to help cattle during seasonal drought conditions. An inexpensive trench silo was demonstrated for farmers to adopt for animal feeding. With reduced acreages, cotton farmers wanted to maximize their yields on the remaining land. Martin Nelson's fertilizer recommendations were increasingly implemented by "progressive farmers."

The Experiment Station was proud of its achievements, especially in cotton. In 1931, the Arkansas Seed Growers Association had successfully lobbied for the Seed Certification Act. By 1933 the State Plant Board began to administer the seed certification law. L.M. Humphrey conducted an extensive crossbreeding program with over 1,600 varieties to develop high-yielding, disease- and wilt-resistant cotton. New cotton varieties 'Arkansas 2088' and 'Acala 31' were introduced. In fact, the Arkansas Station had the most extensive Acala breeding program in the southeast and developed several strains from Acala 5, C5, C7 and C10 and others from Acala 5-37.

In addition, the utilization of 4-10-4 fertilizers improved the tensile strength of cotton; in 1928, 10.5% of Arkansas cotton was less than 7/8 in. in staple length, and in 1932, only 2% was less than 7/8 in. On the other hand, in 1928, only 36.1% of the cotton was 1 in. in length or longer, but in 1932, 54.3% of the crop was 1 in. or longer.

Agronomist C.K. McClelland developed the first soybean, 'Arksoy', adapted specifically for Arkansas. He also conducted research with hybrid corn and oats and helped develop a new strain of lespedeza. (Photo courtesy of Photograph Collection, Special Collections, University of Arkansas Libraries.)

As a result, the Arkansas crop increased its value by $1.6 million in 1929, $1 million in 1930 and $1.6 million in 1931. Fertilizer tests indicated that a fertilizer relatively high in phosphorus, such as 4-17-4, was suitable for the hill sections of the state and a 4-10-4 fertilizer was best for the coastal plain. The Experiment Station believed that nitrate of soda, phosphate and muriate of potash could be mixed together as a substitute for factory fertilizer. When the price of cottonseed was low, cottonseed meal could be utilized as a nitrogen fertilizer. In addition, the cotton spacing experiments that commenced in 1922 concluded that thick stands reduced boll weevil and leaf worm damage.

However, progress brought new problems. The expansion of cotton acreage, the introduction of cotton insects and disease and continuous cropping of cotton with no rotation or soil improvement led to new cotton production problems. Entomologist Dwight Isely worked on the outbreaks of cotton aphid coming as a result of extensive use of calcium arsenate for control of the boll weevil and cotton leaf worm. E.M. Cralley and V.H. Young, working with the Bureau of Plant Industry of the U.S. Department of Agriculture, developed a breeding program to examine the problem of cotton wilt as related to soil nutrient levels, focusing on the combination of high phosphate and low potash levels as a precondition to infection.

The sandy loam soils of the eastern Arkansas Delta on the flood plain of the Mississippi and its tributaries, the St. Francis and the White Rivers, were excellent for cotton, but continuous cropping resulted in soil depletion. In 1937 Martin Nelson's first experiments showed that nitrogen and phosphorus could replenish the soil. R.P. Bartholomew's experiments indicated that sophisticated on-site analysis of soils was necessary for optimal growth. Tests by P. Grissom and W.I. Purgeon showed variable responses to individual nutrients by specific soil types on different locations.

In corn production, the Experiment Station emphasized the importance of cover crops to protect against loss of nitrogen. In order to avoid southern rootworm, bottom lands had to be kept free of wild grasses. Also the Experiment Station devoted large tracts to oats. Seed varieties indicated that two varieties of winter oats showed promise in the state: 'Lee' in the north and 'Curtis' in the south. Spring oats were more difficult because of their susceptibility to hot, dry weather.

The Agricultural Adjustment Administration program of land retirement to raise cotton prices brought accelerated interest on forage, hay and pasture crops. Cowpeas had long been recommended by the Experiment Station. The new soybean variety, 'Arksoy', was to become the eventual basis for a valuable new agricultural sector that would revolutionize eastern Arkansas. To deal with the leaf beetle that attacked the vegetative parts of beans, cowpeas and soybeans, the Experiment Station recommended early June planting to avoid injury. For forage, alfalfa was suggested for planting on well-drained land, cleared of weeds and fertilized with manure or superphosphate. Alfalfa proved to be a successful feed for sows as well as for Holsteins for milk production. Another crop that attracted much interest was grain sorghum as a substitute in drought conditions for less hardy corn.

The Rice Branch Experiment Station at Stuttgart worked with development of better yielding rice varieties, desirable rates and dates of seeding, cultural practices in grass and weed control, rice diseases, possibility of soil improvement and fertilization, irrigation costs of producing rice and other problems of concern to rice growers. Some of these aspects of rice cultivation were more amenable to short-term successes than others. Seed variety and planting time tests yielded applicable results. Roy Adair, a U.S. Department of

Agriculture rice breeder, developed 'Zenith' (1936), which became Arkansas' leading variety. On the other hand, weed control and rice fertilization presented more difficult problems. Conditions under which rice grows also favor aquatic weeds and grasses. A series of experiments conducted over a five-year period revealed that a two-year rotation with soybeans appeared to be the most satisfactory combination. Fertilizer experiments proved to be very frustrating. The usual fertilizers of super phosphate, ammonium sulphate, nitrate of soda, urea and muriate of potash were counterproductive as they not only failed to increase yields of rice, but greatly stimulated the growth of grass and weeds. Instead, to maintain soil fertility, the Experiment Station recommended planting of cover crops such as rye, hairy vetch, crimson clover and Austrian field peas. In addition, tests of soil chemistry indicated that exchangeable potassium, sodium and aluminum ions increase with the depth of the soil while the amount of exchangeable iron, ammonia and manganese decreases. The decline in iron content appeared to increase manganese toxicity to rice plants.

> First Rice Variety
> Developed by the Arkansas
> Agricultural Experiment
> Station
>
> ZENITH, 1936

The horticulture department continued work with apples, peaches, strawberries, grapes and other fruits and vegetables. Research with horticultural crops was aimed at determining the adaptability and desirability of different fruits and vegetables, pest control and marketing. Horticulturists dealt with the decline in Arkansas' apple industry, recommending the use of winter cover of hairy vetch, application of barnyard manure and control of the codling moth by trapping and spraying. They pointed out the importance of destroying overwintering larvae in their cocoons by scraping the rough bark from trees in winter and by destroying the cocoons under trash on the ground or under boards and stones.

The Experiment Station sought to analyze the reasons for the continued decline of the fruit industry. Lack of finances for fertilizers and labor combined with low yields, poor markets, diseases and insect pests discouraged growers. High freight rates placed Arkansas producers at a price disadvantage. In contrast, the Station was fostering a grape industry by grafting 'Campbell Early' onto more vigorous stock. However, cutworms were a major pest for grapes,

Delta Branch Station, formerly called the Delta Substation, was established in 1938 near Clarkedale primarily because it had soil types representative of the region. It was a component of the Cotton Branch Station in Marianna until it became an independent branch station in 1957. Study days such as the one shown above were held often to introduce new technology to farmers.

and farmers were advised to put bands of tanglefoot, an insect tac trap, at the base of the vine to prevent cutworms from climbing up into the vine.

The agricultural chemistry department expanded its studies of the impact of vitamin deficiencies. Barnett Sure conducted studies on the nutritional value of a dietary factor that could be extracted from yellow corn, maize and hempseed. This factor, a necessity for nursing babies, was eventually labelled vitamin E, and wheat oil was discovered to be its best source. The Experiment Station concluded that nutritional diseases of the stomach and intestine associated with diarrhea could be reduced by use of cod liver oil added to orange juice. These discoveries ultimately assisted in reducing infant mortality. Further exploration of vitamin deficiency revealed that anorexia could be cured by concentration of vitamin B, which not only stimulated appetite, but also brought better assimilation of food. M.C. Kik demonstrated that rice polish was an effective cure for pellagra. Margaret Smith informed clients of the importance of eating spinach, carrots, tomatoes and lettuce. The agricultural chemistry department worked with home economists to determine methods of ascertaining vitamin deficiencies by means of hemoglobin determination, red cell counts and refractive index of blood serum.

The election of Governor Carl Bailey in 1936 brought a renewed vigor in state government. The Experiment Station worked with Bailey to make state government more efficient. C.O. Brannen in 1937 came out with a plan for consolidated county government that would lower costs and result in a more efficient management system (Bulletin 338). Brannen believed that the quorum court had to be replaced by sophisticated, university-trained county managers who could better deploy existing financial resources without political consideration. However, vested political interests thwarted the implementation of his suggestion.

The Experiment Station broadened the scope of its research in 1936 to include tests of ornamental plants. J.R. Cooper started plantings of azaleas, junipers, roses and evergreens to ascertain their hardiness and drought resistance. J.E. Vaile tested varieties of grapes, strawberries and blackberries. In forestry, R.D. Steven began an analysis of the distribution of tree varieties in Arkansas.

Assistance for the rice industry continued to be a major research effort. Diseases of rice were studied by E.M. Cralley and E.C. Tullis in an effort to eliminate stem rot by hybrid selection. In addition, the scientists discovered that changes in soil chemistry as a result of losses of acidity led to chlorosis. Research conducted at the Experiment Station revealed that the addition of sulphur contributed to increased rice yields. Moreover, the rice station studied the control of stem rot of rice caused by sclerotia, which was a serious disease often causing the loss of bushels per acre. The technique developed involved spacing of irrigation periods.

In 1938, Congress passed the second Agricultural Adjustment Act. The new comprehensive approach to persistent surpluses of staple commodities such as corn, cotton and rice called for establishment of regional laboratories to work with the U.S. Department of Agriculture and state Experiment Stations to promote new varieties and to examine new marketing techniques. The act also called for acreage restrictions based on the previous year's production to promote conservation. The impact of this program accelerated the flow of tenants from cotton lands. As one landowner related, "I took the government payments, brought tractors and got rid of my croppers." Landowners had an incentive to buy fertilizers and improved seed varieties. Farmers who owned their own land realized a triple return: governmental allotments, higher prices for cotton and alternative crops.

The cotton program remained the single most important program of the Station. In 1936, L.M. Humphrey had started a program of inbreeding cotton to develop a uniform staple length and lint percentage to help in the difficult, time-consuming task of measuring the length of cotton fibers. Burt Johnson in 1939 developed an apparatus that considerably shortened the task (Bulletin 381). First a mechanical comb parallelized the fibers still attached to the seed. Then a multiple-seed photo-electric sorter separated the fibers of seed cotton into their respective length groups. In 1938 V.H. Young's tests on control of cotton wilt and "rust" concluded that the combination of a moderately wilt-resistant variety of cotton and use of a fertilizer with potash resulted in a marked reduction of cotton wilt (Bulletin 358); however, the problem of root knot was still present.

In 1938, Lewis Turner began pioneer studies of the pine growing potential of the coastal plain region with the objective of making detailed correlations of slope, fertility and drainage to help revive the depressed timber industry. The federal government's rural rehabilitation program had as its objective the relocating of farmers off "submarginal" land and the conversion of these areas from row crops to either forest or pasture. The Experiment Station would study various techniques to successfully facilitate this transition to more viable economic pursuits.

In 1939, J.G. McNeely, rural sociologist, assisted by the Agricultural Adjustment Administration and United States Bureau of Agricultural Economics, made an in-depth study of the impact of the New Deal on Arkansas farmers. Three counties, Jefferson, Miller and Phillips, were selected as being geographically representative, but other counties were also surveyed. The statistics for displaced farmers ranged from 25% in Pope County and 20% in Chicot County to 7% in Clark County as farmers substituted day labor for tenants. The studies related that many of the displaced tenants moved to nearby agricultural communities in search of relief work. These tenants, without money or land, actually were worse off than the tenants of the previous decade.

W.H. Metzler, the demographic expert on the Experiment Station, was pessimistic about the future of large numbers of rural Arkansans. He reported that "poverty, bad living conditions, antisocial attitudes, and socially maladjusted individuals are likely to increase." Soil erosion, deforestation, lack of fertilizers and persistent drought had undermined all hope of progress. In 1941, Isabella

During the late 1930s, forester Lewis Turner conducted pioneering work to help revive the depressed timber industry in the costal plains. Station researchers studied ways to convert some of the marginal row crop land to forest or pasture. (Photo courtesy Frederick B. Turner.)

Wilson in the department of home economics completed a comprehensive five-year study of health conditions of rural Arkansans. Her results mirrored the studies of rural sociologists. Poverty, ignorance, malnutrition and lack of available doctors contributed to high infant mortality and a pervasive poor health pattern among rural Arkansans.

The Experiment Station tried to demonstrate ways to alleviate some of the problems of the rural poor. Barnett Sure and M.C. Kik's research indicated that the addition of rice polish containing vitamin B and B_6 complex to diet could result in a major improvement in chronic rheumatism and heart disease. As part of his efforts with the Rural Rehabilitation Agency, Deane G. Carter in agricultural engineering presented house plans that could make adequate shelter readily available at a low cost with some assistance.

Agricultural scientists started worrying about the ecological deterioration of Arkansas agriculture and problems in rice culture. The cyclical droughts of the 1930s contributed to a lowering of the water table in the rice-growing region. Kyle Engler proposed the use of reservoirs and channelized water to supplement diminished subsurface water. Orville J. Hall, rural economist, tried to assess the problems of rice farmers. Hall studied the method and time of the sale of rough rice as it correlated with farmers' forecasts of rice prices in the months following the threshing of the rice. Rice prices at threshing time had to be balanced against the alternative costs of handling and storing and the estimated price at some future time.

In contrast to the gloomy assessment of difficulties in rice, one of the few bright spots in agriculture was the growth of the poultry industry. S.R. Johnson and R.M. Smith started experiments to determine an economical but nutritious chicken ration. They were interested in developing a cheap and readily available feed that would lower costs to allow Arkansas farmers to receive a reasonable profit. W.R. Horlacher worked with crossbreeding of chickens in an effort to produce a vigorous breed by combining Rhode Island Red females with White Wyandotte males. As part of the breeding program in 1941, crossbreeding produced a chicken that required 13.33% less feed. By 1942, growers were utilizing the new breed of White Wyandotte x New Hampshire.

One persistent problem of Arkansas farmers was the low yield for corn. R.P. Bartholomew, working at the Experiment Station farm, tried to demonstrate to farmers that the use of hybrids, crop rotation and adequate fertilizer could more than double yield per acre. In particular, Bartholomew tried to show that corn yields could be doubled if Experiment Station guidelines were followed, especially regarding the applicability of irrigation.

Crop failures in 1930, 1934 and 1936 had slowed down his efforts, but by 1937, C.K. McClelland had begun an expanded corn program

W. R. Horlacher, head of the department of animal industry, became the eighth director of the Arkansas Agricultural Experiment Station, succeeding Dan Gray in 1939. Horlacher resigned in 1946 to accept a position at Colorado State College.

involving some 3,000 lines. By 1941, some 700 remained that had demonstrated favorable characteristics. McClelland's goal was to give Arkansas farmers both yellow and white hybrids of superior yielding ability, fortified with disease, insect and drought resistance and the ability to withstand strong winds.

In 1939, W.R. Horlacher succeeded Dan Gray as director of the Station with C.O. Brannen remaining as assistant director. Arkansas was on the verge of the scientific revolution for which the Experiment Station had been laying the basis for 52 years. Despite the important work in fertilizers, soil erosion, insect and pest control and feed analysis, the lack of capital, depressed prices for commodities and lack of education interfered with implementation of the results.

By 1940, Arkansas agriculture along with that of the rest of the South showed some faint signs of revival as a consequence of New Deal policies. Nobody could envisage the whirlwind of change that was to engulf American agriculture with the onset of World War II and the accompanying scientific and technological revolution.

Chapter V

The Scientific Revolution 1941-1960

In the wake of the economic and social devastation caused by the Great Depression came the onset of World War II and the agricultural revolution long predicted by scientists. The continued decline of the farm population accompanied by farm consolidation allowed for the economies of size necessary to justify application of the various cultivation techniques, better seed varieties and fertilizers. The higher prices engendered by wartime inflation and labor shortages necessitated the extensive use of labor-saving machinery. In order to utilize more fully the resources of the state Agricultural Experiment Stations, the U.S. Department of Agriculture formed the Agricultural Research Administration in 1941, combining the office of Experiment Stations with the scientific and research entities of the Department of Agriculture in order to facilitate closer cooperation between federal and state research programs. However, the individual state Experiment Stations retained their autonomy.

The state Experiment Stations faced a unique challenge. Researchers were told to find ways to strengthen existing food stocks and to find substitutes for products such as oil in the face of shortages of fertilizers, seeds, petroleum and insecticides. The movement toward agricultural diversification was accelerated by the need to feed not only the military but also the Allies. Waste had to be minimized by improvements in food preservation. In spite of the dramatic rise in cotton prices from 10 cents/lb in 1939 to 22 cents/lb in 1945, farmers had the incentive to grow soybeans, peanuts, fruits and vegetables and raise cattle. Marginal operators and tenants had the economic alternative of factory jobs.

The Experiment Station had been in the forefront of implementation of programs for the revitalization of rural Arkansas. Agricultural engineer Deane G. Carter, who had been making house plans for rural rehabilitation, turned his attention to model fields to prevent erosion as part of the land restoration program of the Soil Conservation Service. His colleague, Kyle Engler, worked with

Agricultural engineer Kyle Engler (second from left, first row) conducted studies to conserve underground water supplies as the droughts of the 1930s seriously affected crop production. Xzin McNeal (first row on left) conducted research on artificial drying of combined rice, and in the l960s he developed the first machine used to stack seed cotton on a pallet. Other members in the department of agricultural engineering in 1950 were (l to r) first row, McNeal, Engler and B.Y. Richardson; second row, H. S. Stanton. T. E. Hazen, R. H. Benedict and G. W. Steinbruegge.

farmers in southeastern Arkansas in combating the decline of underground water levels. The persistent, recurrent droughts of the 1930s had badly depleted water resources, resulting in an interest in irrigation. Engler worked with farmers to draft water rights legislation as well as to suggest alternative sources of water.

The higher prices for beef presented opportunities for the growth of the cattle industry. In addition, acres diverted from horses and dairy cattle were available for production of forage. Addressing the problems faced by Arkansas stock grazers, the first federally funded project on performance testing of beef cattle was begun. Warren Gifford in animal industry felt that malnutrition, parasites and paratuberculosis along with a lack of forage and pasture management were hurting development of the cattle industry. M.C. Kik advocated the use of Atlas sorgo silage, which had been brought to the United States from Africa and was both drought resistant and an excellent fodder, to deal with a forage problem exacerbated by a scarcity of corn. However, the plant could poison livestock if fed in certain stages of development.

The creation of a viable beef cattle industry had been impeded by the absence of alternatives to corn, which did not thrive in Arkansas.

H.R. Rosen started a long-term project on crossbreeding varieties of oats to develop an oat resistant to crown rust. After a decade of experiments, 'DeSoto' was released (1942), later followed by 'Traveler' (1944), varieties that were resistant to smut as well as being high yielding and winter hardy and providing a high-quality forage. Oats had not flourished in Arkansas due to fungus attacks, erratic weather and winter kills.

Interest in nutrition grew during the scarcity of the war years. Barnett Sure's long-term work in vitamin deficiency resulted in his discovery of the efficacy of liver extract in dealing with vitamin deficiencies. The population of rural Arkansas had long been characterized by major nutritional deficiencies as the result of inadequate diet. Sure believed that the high infant mortality rate could be reduced by vitamin supplementation.

Experiment Station scientists continued the never-ending struggle with plant diseases. S.B. Locke of the department of plant pathology attacked tomato blight by crossbreeding domestic tomatoes with wild tomatoes having natural blight resistance in an ultimately successful effort to create a natural resistance. The use of non-chemical controls was also utilized by E.M. Cralley, who helped demonstrate that low potassium combined with nitrogen and phosphorus could check cotton fungus. V.H. Young found that cotton blight could be controlled by treating cottonseed with mercury chloride or ethylmercury phosphate dust. Dwight Isely continued to preach to farmers that cotton pests could be controlled by judicious use, or "spot dusting," of calcium arsenate and winter clearing of the hibernation places of boll weevils. W.J. Baerg helped develop a compound to destroy fall webworm caterpillars that attacked pecan, walnut, persimmon and hickory trees.

As the war proceeded, it took its toll on Experiment Station personnel. Thirty percent of the Experiment Station staff had joined the armed forces, and agricultural engineer A.H. Thompson was killed in action. The federal and state governments could not give additional financial support to the Experiment Station despite the additional burdens the Station faced. This reduction in force was a microcosm paralleling developments in Arkansas' agricultural sector where fewer tenants and farmers had to increase production. C.O. Brannen reported that the tenant population had been reduced some 60% from the 1932 figure.

The search for labor-saving devices was accelerated in the face of labor shortages. Agricultural engineer Xzin McNeal started work on

Plant pathologists S. B. Locke, E. M. Cralley and V. H. Young (Young is pictured on right with D. B.Shank, left) evaluated new strategies to control plant diseases through genetics and biological controls. (Photo courtesy Photograph Collection, Special Collections, University of Arkansas Libraries.)

an experimental rice drier. The forced hot air drier was to take years to perfect, but it would revolutionize the harvesting of rice. The problem of the impact of drought on animal forage was countered by an expansion in drought-resistant sorghum. Cotton farmers were able to increase yields by approximately 25% by introducing new varieties recommended by the Station as well as by following fertilizer application guidelines.

The war years with their accompanying shortages made scientists seek alternative means of feeding people and animals. Animal protein, meat scraps, fish meal and skim milk were all in short supply. S.R. Johnson of the department of animal industry emphasized that pasture improvement and new breeds were essential to assist farmers in meeting the demand. The department worked with purebred Poland China hogs and Aberdeen Angus and Hereford cattle.

With the greater need for meat, northwest Arkansas agriculture began to shift its primary thrust from apple orchards to broilers. Production of chickens increased from 3 million in 1936 to 15 million in 1942. To assist growers, R.M. Smith in the department of animal industry crossbred New Hampshire and Barred Plymouth Rock chickens to produce chickens with greater meat production. Smith and S.R. Johnson worked on low-cost chicken feed.

The use of nitrates in high explosives led to a fertilizer shortage that forced scientists to seek alternatives sources of soil enrichment, a necessity in the heavily depleted soils of Arkansas. Consequently, agricultural scientists renewed emphasis on green manure as a substitute for chemicals. R.P. Bartholomew, working with the National Agricultural Conservation Program, emphasized that hairy vetch, crimson clover, burr clover and Austrian winter pea were excellent winter covers. Bartholomew also suggested the use of nitrate of soda and urea as a substitute fertilizer.

The need for food preservation for soldiers was a major war priority. The Experiment Station built a dehydration laboratory. V.M. Watts believed that initial experiments demonstrated that many dehydrated vegetables would rot and were not acceptable to the popular taste. Barnett Sure tested for the vitamin content of the new dehydrated products.

The war also encouraged development of new crops such as peanuts and soybeans in Arkansas. The loss of palm oil as a result of the Japanese conquest of Southeast Asia stimulated the search for alternative sources of oil. Prior to the war, cottonseed oil, soybean oil and imported vegetable oils all suffered from a discriminatory tax to discourage sales of oleomargarine. However, during the war, governmental policy in allocating rationing points favored the purchase of margarine over that of butter by allowing the transfer of points for more expensive cuts of meat. After the war, dairy manufacturers shifted to ice cream, cheese and fluid milk as alternatives to butter production as the new purchasing pattern persisted in the consumer mentality.

By the end of World War II, the Arkansas state legislature had purchased an additional 185 acres for the Main Experiment Station due to the proliferation of new crops. Fears of a return to the economic conditions of the 1930s spurred an interest in expanded support for agricultural research. Lines of research commenced during the war continued during the post-war period. E.M. Cralley tested new techniques for treatment of corn seed prior to planting.

V. M. Watts, head of the department of horticulture and forestry, conducted studies on food preservation techniques, dehydrated vegetables and tomato breeding. He also developed a new watermelon variety called the 'Hope Diamond'.

L.C. Kapp and R.P. Bartholomew continued exploring efficient methods of irrigation needed to prepare seed beds and control weeds for cotton. M.C. Kik tested the impact of dehydration on the vitamin content and nutritional value of vegetables. W.J. Baerg continued his interest in biological control and pest management. Martin Nelson sought to assist the growing beef industry by pasture studies emphasizing the need for adequate fertilizer and measured cutting of hay.

In 1945, the United States government spearheaded the United Nations Relief Rehabilitation Administration to distribute food in Europe. Post-war droughts in Europe in 1946 and 1947 continued to sustain demand for American foodstuffs. Furthermore, the Marshall Plan, enacted in 1948, provided the Europeans credit to purchase American agricultural products. Virginia Congressman John Flannagan proposed that the state Experiment Stations be given additional money to discover new uses for farm products. Kansas Congressman Clifford Hope wanted new funding to be used to research marketing and distribution of agricultural commodities. The two approaches were merged in the Research and Marketing Act of 1946, which called for a gradual increase in appropriations of $2.5 million for the state Experiment Stations. In order to deal with anticipated commodity surpluses, 20% of the money was divided to

M.C. Kik served in the department of agricultural chemistry from 1927 to 1967. His research focused on the nutritive value of milk proteins and of rice and its byproducts. He developed the process of parboiling rice.

the states on a matching basis for marketing, packaging, transportation, storage, grading and distribution studies. The rest of the funds could be utilized for expanded research and increased support of the regional laboratories and cooperative regional projects.

Exacerbating fears of a farm crisis in the post-war era was the demographic trend already underway: a population shifting from rural to urban areas. The wartime emphasis on raising production levels accelerated the decline in farm tenancy and its accompanying social as well as economic implications. To monitor these developments, C.O. Brannen was made the chairman of the Southwest Land Tenure Research Committee in 1947. The regional study explored the impact of technology and market changes on rural America. In particular, Brannen was concerned with the alternative industrial opportunities that would have to be created to assist displaced tenants.

By 1946 the mechanization of cotton picking by the Rust cotton picker furthered the elimination of hand labor. Employment of new techniques brought new challenges to the agricultural scientist. Mechanical picking dramatically shortened the picking time; however, the higher moisture content and increased trash and leaves in the mechanically picked cotton required new drying and cleaning equipment. Experiment Station scientists responded by developing 'Arkot 2-1' cotton in cooperation with the U.S. Department of Agriculture, a cotton more suitable for mechanical picking. Agricultural engineer Xzin McNeal worked on methods of storage of cotton and cottonseed, and the first flame weed killers and airplane spraying of defoliants were studied.

In 1946, Lippert S. Ellis was appointed director of the Agricultural Cooperative Extension Service and Agricultural Experiment Station as well as dean of the College of Agriculture. Ellis had received his degrees from the University of Wisconsin and had been employed at Oklahoma A&M College and by the U.S. Department of Agriculture prior to his elevation to the directorship.

In rice, Xzin McNeal and Kyle Engler made an important contribution to mechanized harvesting by improvements to the continuous flow rice drier, allowing for more rapid handling of the increasingly mechanized rice harvest. In coordination with these efforts, Barnett Sure worked on new methods of parboiling rice to shorten the cooking time, making rice more acceptable to the consumer while ensuring it retained its vitamin content. Moreover, the animal industry department demonstrated that rice bran and polish could be fed to poultry if supplemented by Vitamin A.

In 1946 Lippert S. Ellis was appointed director of the Station and the Cooperative Extension Service and dean of the College of Agriculture.

The Experiment Station's long-term efforts at development of new varieties became even more important. New varieties of watermelons, strawberries, grapes, peaches and tomatoes were introduced. In 1947, 'Rexark' was released as a new rice variety.

The beef and dairy industry in northeast Arkansas, which had gradually increased in the 1930s as the government retired submarginal agricultural land, received a major boost as a result of continuing high post-war demand. The Experiment Station started a pasture-improvement program to develop guidelines to assist farmers. The Batesville station had major success filling and covering gullies with woven wire and Bermudagrass sod as preparation for the creation of pasture. In addition, the Station developed techniques to aid farmers in culling low producing cows. However, problems of breeding, feeding and management of diseases still remained.

The feared post-war collapse in prices had not occurred. Instead these years witnessed the beginning of a long-awaited technological revolution in Arkansas agriculture. Lippert Ellis viewed the results as being "clearly seen on every hand throughout the state." To promote agricultural progress, the Experiment Station created a series of substations on leased land.

The Peach Substation was established in 1948 on 45 1/2 acres northwest of Nashville under the direction of Gene Arrington. The first varieties of Elberta peaches on the Experiment Station were planted here. Tests included cover crops, spacing, pruning, thinning, rest period, fertilization, spraying, winter injury, growth hormones, breeding, root stock, irrigation and marketing. Also studied were fruit moth and peach tree borer as well as control of brown rot, scab and bacterial spots. An inexpensive weed control program was initiated. Irrigation tests were started, and a drip irrigation system was developed. One project was the evaluation of several varieties of peaches budded on a number of different rootstocks. In the spring of 1987, Gene Arrington retired, and the peach substation was phased out.

The Strawberry Substation was established in early 1948 under the direction of A.T. McDaniel on 20 acres of leased land in Bald Knob. Until 1960, research on the substation was primarily in strawberries with work on blackberries and okra. Researchers worked with the Birdseye Division of General Foods Corporation and Atkins Pickles to test vegetable crops of okra, summer squash, southernpeas, southern greens and cucumber. In 1965, the work

Gene Arrington directed the research activities at the Peach Substation near Nashville from its opening in 1948 until it ceased operation in 1987.

was expanded to incorporate additional work in vegetables, blackberries, blueberries, raspberries, muscadine grapes and egg plant. The program was directed at developing fresh market cultivars with disease and insect resistance suitable for mechanical harvesting and processing. Arkansas strawberry growers utilized chemical weed control, disease and insect control. New hybrids of cucumber, summer squash and summer peas were introduced. Studies commenced on southern greens and okra relating to fertilization, spacing, insects and disease. In 1966 the station moved to a new 76-acre site five miles north of Bald Knob.

In 1959 the Fruit Substation was relocated 7 miles north of Clarksville, eventually encompassing 230 acres. The substation was originally established in 1948 near Lamar with Cole Westbrook in charge but was moved to Clarksville when part of the property was lost to the construction of Interstate 40. The program originally involved peaches but was expanded to include apples, pears, plums, nectarines, apricots, cherries, grapes, blueberries and blackberries. The blackberry breeding program has become the largest in the world. An apple breeding program to develop early-season cutting was undertaken. Today, cultivar evaluation is conducted at the substation on apples, apricots, blackberries, blueberries, cherries, grapes, peaches, pears, plums and nectarines. Cultural research includes plant nutrition, training and pruning. The evaluation of pesticides for fruit crops includes testing of fungicides, herbicides and insecticides.

The post-war triple revolution of chemicals, varieties and mechanization seemed to promise what appeared to be amazing results.

The Fruit Substation, established in 1948 near Lamar and moved in 1959 to Clarksville, has the largest blackberry breeding program in the world and the second largest table grape breeding program in North America.

Greater use of tractors, pesticides and insecticides contributed to the reshaping of Arkansas agriculture. However, Lippert Ellis warned against any scientific quick fixes, saying, "research is a slow, costly process; accomplishments don't come overnight. But as a problem is solved, the answer goes out to the state's farmers."

From the inception of the Experiment Station, the dissemination of information was considered a major function; the Station bulletin series had begun in 1886. With the rapid changes affecting all facets of agriculture, the Experiment Station added the report series in 1945, the mimeograph series in 1949 (now known as the research series) and *Arkansas Farm Research* in 1951. The first mimeographs dealt with the importance of bees; applications, limitations and dangers of 2,4-D and 2,4,5-T; problems of mechanized cotton pickers; and the preparation of seed beds and planting of cotton. The first report series publications were on yield tests of corn, rice and oats. The special report series was started in 1957.

Post-war mechanization resulted in the continued exodus of tenants. In 1947 Erling Solberg succeeded C.O. Brannen in the Southwestern Land Tenure Research Committee, composed of representatives from Arkansas, Louisiana, Mississippi, Oklahoma, Texas and the U.S. Department of Agriculture. The committee wrote a legal manual for landlord-tenant statutes to regularize the rights and obligations of tenants, solving a major problem that had long plagued agricultural production. However, the underlying causes of tenant flight, mechanization and the lure of better opportunities outside the South, continued to persist. The Truman administration reimposed rigid acreage restrictions in order to forestall surpluses, reinforcing the rural exodus and accelerating the consolidation of smaller, marginal farmsteads.

The Experiment Station cotton tests conducted at the Cotton Branch Station at Marianna and at a temporary Alfalfa Substation at Osceola tested for planting dates, fertilizer application rates and varietal influence under the direction of Charles Hughes in the department of agronomy. In 1956, due to the impact of the Soil Bank program, the experimental tests emphasized greater efficiency in production.

The introduction of new insecticides seemed to promise the long-sought control of the boll weevil and cotton aphid. The mechanization of cotton picking combined with chemical defoliation of cotton changed cultivation practices to narrow rows and flame destruction of weeds. Chemical defoliation was a necessary preparation for mechanical picking in order to reduce the trash that lowered the cotton grade. By 1963, 72% of the cotton harvest was machine picked. Strains of cotton were developed that ripened at a uniform time, thus reducing the number of times that pickers had to go through the field. The cotton testing program indicated that 'Delfos 9169' and 'Stoneville 2-B' were adaptable to mechanical picking.

The scientific revolution in medicine as a result of antibiotics spread to the treatment of animal diseases in the post-war period. Penicillin and sulfa medicines were applied to animal diseases. Edward Cook found that streptococcus mastitis in bovines could be cured. Robert Smith concluded that antibiotics along with commercial vitamins and high proteins improved growth and health of chickens.

The growth of the poultry industry in Arkansas was paralleled by similar developments in Maryland and Georgia. As a consequence, producers were locked in a highly competitive situation in which cost efficiency meant the difference between profit and loss. The constant pressure to reduce costs by cross-breeding chickens and utilizing inexpensive, high-protein rations stimulated poultry science in Arkansas.

The higher income levels of the post-war period, better highway transportation and the need to diversify all kindled a renewed emphasis on fruits and vegetables, especially for the small farmer who could not afford to mechanize his operations. The revival of the fruit and vegetable industry in Arkansas after the doldrums of the Great Depression proved to be a difficult task in the face of disease, insect infestation, soil erosion and loss of traditional Midwestern markets. The Experiment Station leased farmsteads to launch studies of watermelons, peaches, strawberries, alfalfa, squash, tomatoes and sweet corn, working cooperatively with agents of the U.S. Department of Agriculture. In particular, in view of the recurrent droughts that had afflicted Arkansas agriculture, irrigation techniques were studied along with the traditional areas of fertilizers, cultivation practices and disease and insect control.

The Experiment Station broadened its soil testing service, which was first offered in 1945 at Gray Hall on the Fayetteville campus and later moved to a renovated mule barn on the farm. The Legislature in 1953 enacted Act 301 placing a fee of 37.5 cents per ton on fertilizer with 25 cents to go to the University for soil testing and research. This increased financial support enabled the Station to begin operations at the Eastern Arkansas Soil Testing Laboratory at Marianna in 1954 and provided the basis for the cotton nitrate monitoring program. Between 1954 and 1988, 1.8 million soil samples were tested by the Station. Glenn W. Hardy, who served as dean of the College of Agriculture and Home Economics from 1965 until 1987, was one of the first agronomists to work in the laboratory.

A major contribution to the study of pesticide application was made by Fred E. Whitehead, who developed in 1951 a new method of application of insecticides by use of a granular carrier rather than by dust or spray. The innovation was to be widely adapted as the preferred method of application for many kinds of pests due to the ability of the granules to penetrate the plant canopy. Whitehead developed equipment, dosage levels and sizing, drying and storage techniques of the granular carrier.

On an ominous note, Station entomologists noted the possible deleterious nature of DDT. In 1939 a Swiss company had formulated the insecticide DDT, and by 1943 it was introduced in the United States. In 1944, W.J. Baerg started tests of DDT in Arkansas. Baerg believed that indiscriminant use of the insecticide could be dangerous and emphasized that biological control was preferable. Careful testing indicated what was to be widely recognized in the following decade: extensive use of DDT badly upset the balance of nature. In a foreshadowing of later studies, Station scientists concluded that insecticides could pose an ecological danger.

The Experiment Station was still concerned about rural emigration and changes in land tenure. In 1950, the Experiment Station participated in the regional analysis of the changing face of Arkansas land tenure. The decline in commodity prices, increased mechanization and limited access to capital resulted in a consolidation of land in Arkansas into more economically viable holdings.

The return of the problem of unsold agricultural surpluses caused the federal government to shift direction to more flexible, lower price supports. The first attempts to lift acreage controls on rice accelerated the movement toward greater mechanization of harvest; consequently, a surplus developed in 1949 as normal rice production in Asia resumed, closing Arkansas rice producers' most lucrative markets. In response to the anticipated build up in unsold rice, the Truman administration froze the amount of new rice acreage that could be brought into production. Some producers started to switch their land to soybean production, which was still unrestricted. Oth-

Fred E. Whitehead investigated new methods of insecticide application but was one of the first researchers to emphasize the dangers of pesticides to the environment. He developed the process of using granular particles as carriers of insecticides. The pesticide industry adapted this procedure of insecticide application, and now granules are used to control insects on farms, gardens and lawns. Unfortunately for the Station, the process was not patented.

ers started to use 'Rexark' rice to get higher yields in the land still in rice production. The full impact of the dramatic shift in government programs was mitigated by the outbreak of the Korean War and renewed demand for rice. In the next four years, advances in seeding, fertilizer, air drying and use of combines led to production increases from 2.2 million to 3.6 million barrels (1 barrel=162 lb). Consequently, reimposition of the allotment of rice acreage in 1955 as part of President Eisenhower's Soil Bank Program fell even harder on rice farmers. To slow the expected buildup of rice surpluses, one quarter of all existing rice land was to be taken out of rice farming and converted to another nonrestricted crop or pasture.

In this period of shift from a period of agricultural scarcity to oversupply, Congressman Stephen Pace of Georgia called for an even greater degree of coordination among state Experiment Stations, regional laboratories and the U.S. Department of Agriculture to avoid duplication of effort. As a result of his investigative hearings in 1953, the Agricultural Research Service was created.

With the diversification of agricultural production, the Arkansas Station was to emerge as a major catalyst in the rapid expansion of poultry production. The stimulus to this interest in chicken production had been the shortage of beef during World War II. The Ozarks had witnessed a decade of growth in the poultry industry with commercial broilers increasing 201% from 8 million birds in 1939 to 24

VARIETIES AND STRAINS DEVELOPED BY THE ARKANSAS AGRICULTURAL EXPERIMENT STATION 1941-1960

AGRONOMIC CROPS	
Cotton	
NR-AHA	1955
Rex	1957
Oats	
DeSoto	1942
Traveler	1944
Delair	1949
Taggart	1948
Arkwin	1952
Ferguson 560	1956
Rice	
Arkrose	1942
Prelude	1943
Kamrose	1943
Nira 43	1943
Arkansas Fortuna	1943
Rexark	1947
Wheat	
Ace	1960
HORTICULTURAL CROPS	
Southern Pea	
Monarch Blackeye	1957
Tomato	
Fortune	1949
Indark	1954
Watermelon	
Whitehope	1955
Hope Diamond	1958

million in 1949 to 78.5 million by 1954. Broiler production expanded rapidly in Arkansas with a complete shift from independent to contract growing in which growers supplied buildings, equipment, utilities and labor and the processor supplied feed, chicks and field service.

In 1951, the Arkansas Agricultural Experiment Station hosted the Chicken of Tomorrow contest, which greatly stimulated both poultry research and the poultry industry in Arkansas. Howard Pierce, the president of the Great Atlantic and Pacific Tea Company, the single largest retailer of food in America, had called for poultry breeders to develop a broad-breasted chicken that would have the white meat demanded by consumers. In order to bring poultry producers' interest to a peak, the A&P, working with the U.S. Department of Agriculture, sponsored the national contest with cash prizes and accompanying publicity for the winner. The Arkansas Experiment Station, with the assistance of leading poultry producers John Tyson and Roy Ritter of Springdale, agreed to assist in the hosting of the competition. Poultry scientist Robert Smith personally supervised the shipment of eggs and development of the birds. New incubators were installed at the Experiment Station. The Kansas, Missouri and Oklahoma Experiment Stations lent their assistance to the Arkansas Experiment Station for a twelve-week period. The chickens were weighed at the pens by special crews. The Experiment Station's poultry experts helped in judging the various categories of hatchability, fertility, livability, feeding ability, weight, rate of feathering and uniformity of type and color.

To symbolize the significance of the event, United States Vice President Alben Barkley gave the $5,000 award to the winner at a ceremony held in Razorback Stadium before a crowd of approximately 8,000 people during the concluding week of festivities. Some 30,000 people reviewed the various exhibits. The impact of the contest on the broiler industry was to stimulate interest in improved breeding and to pave the way for Arkansas producers to become leaders in poultry.

To assist the incipient commercial beef industry, L.H. Hileman and R.L. Beacher in agronomy started testing the response of forage grasses and legumes to fertilizer application. The idea was to promote forage production on permanent pasture essential to the cattle industry. At the Main Station in Fayetteville, construction was begun on an animal science building with two wings completed in 1957 and three more in 1962. The building was one of the few on campus at the time funded by appropriations from the State Legislature.

The Arkansas Agricultural Experiment Station hosted the National Chicken of Tomorrow Contest in 1951. This event helped promote the development of the poultry industry in northwest Arkansas. In the photograph above, members of the Board of Directors of the contest meet in regular session. Shown on left and then clockwise around the table are Jeff D. Brown and Luther Johnson of Springdale; Vick Will, Rogers; Roy C. Ritter, chairman, and John Tyson, both of Springdale; Lloyd Peterson, Decatur; Robert M. Smith, Station researcher; Kirk Hale, Fayetteville; and Charles Garrett, Rogers.

Charles Vantress (left), of the Vantress Poultry Breeding Farm, Live Oak, Calif., receives congratulations of U.S. Vice-President Alben W. Barkley after being awarded the $5,000 first prize in the 1951 National Chicken-of-Tomorrow Contest. Eight thousand people witnessed the presentations in the University of Arkansas' Razorback Stadium.

The early 1950s ushered in another dry cycle similar to that of the 1930s. Ellis told farmers of the need for supplemental irrigation, emphasizing that, despite unfavorable weather patterns, corn yields had doubled, corn and oats had risen by a third and soybeans by 60%. The use of pre- and post-emergent herbicides eliminated tedious, labor-intensive hoeing of weeds. To aid farmers cultivating forage, H.R. Rosen and J.O. York released a new variety of winter oats, 'Arkwin', that had high resistance to the blight that previously had destroyed almost 70% of the winter oats in Arkansas.

The drought cycle of the 1950s resulted in more extensive examination of problems and opportunities afforded by irrigation. Arthur E. Spooner and Charles E. Caviness commenced studies on irrigated cotton, as too much moisture was believed to cause excessive vegetative growth, boll-rot and aggravated shedding.

In the 1950s, long-term research projects of the Station were yielding important advances. By 1952 hybrid corn had become an important part of Arkansas agriculture with 65% of the acreage planted in hybrids. The Arkansas State Plant Board determined which hybrids could be sold on the basis of the corn performance tests.

The challenge of plant diseases was seemingly unrelenting. Pests and fungi continually threatened Arkansas crops. In 1953, Charles Lincoln reported that the pink boll worm, which had already infested Texas cotton, had spread to Arkansas. The Southwestern corn borer was also becoming a major pest. Moreover, house flies had developed an immunity to DDT.

The immediate impact of scientific agriculture was being seen throughout the state. Associate director John White estimated that one third of the total cash receipts derived from Arkansas agriculture could be attributed to the Experiment Station. He predicted that if losses from diseases and insects could be reduced, the declining net income of farmers could be reversed.

After Dwight Eisenhower was elected as President in 1952, Ezra Taft Benson became the Secretary of Agriculture. Benson, an advocate of the free market system, made the reduction of agricultural surpluses his major objective. The passage of the Soil Bank Act in 1955 accelerated the transformation of Arkansas agriculture. The Allotment Act resulted in total cotton acreage being half of what it had been in 1929. In the uplands area, cotton practically disappeared. The introduction of tractors reduced the demand for corn as feed for work animals; consequently, corn acreage fell by two-thirds of the 1929 acreage. Concomitant with these developments, land devoted to beef and dairy cattle doubled. On the abandoned cotton land, farmers grew soybeans and rice.

At the Experiment Station, projects such as the training of cotton scouts to assist farmers in decisions when to apply insecticide were expanded. Arkansas cotton growers were under pressure from California growers who had a competitive advantage since cotton in the irrigated, desert-like conditions of the central valley of California was relatively free from disease and pests. The continually low cotton prices resulted in pressure to mechanize. B.A. Waddle and C.A. Moosberg of the U.S. Department of Agriculture developed 'Rex' cotton in 1957, which was particularly adapted to machine harvesting, matured early and was resistant to angular leaf spot disease and to fusarium wilt as well as being storm resistant.

Entomologist Charles Lincoln (far right, center row) was one of the pioneers in cotton insect management. He recommended the use of insecticides only when insect scouting showed the need for applications. In Station Bulletin 829, he outlined new procedures for scouting and monitoring cotton insects that were used by entomologists throughout the South. Other members of the department of entomology in the late 1950s were (l to r) front row, J.L. Lancaster, Willard Whitcomb, W.J. Baerg (emeritus); second row, L.O. Warren, Robert C. Hunter, Larry Rolston, Floyd D. Miner, Lincoln; back row, Paul Boyer, C.E. McCoy, Gordon Barnes. (Photo courtesy L.O. Warren.)

Brad Waddle (right), the first faculty member to hold the Ben J. Altheimer Chair for Cotton Research in 1951, helped to developed several varieties of cotton, including 'Rex' cotton in 1957. Assisting Waddle is Charles Hughes.

From 1929 to 1959, yields in cotton increased 40% from 152 to 212 lb lint/acre as result of improved production techniques, selection of better adapted and improved varieties and heavier use of fertilizer. By the 1950s, most farmers purchased seed certified by the Plant Board and treated with Ceresan. Varieties of wilt-resistant 'Arkot 2-1' were popular.

The utilization of the mechanical cotton picker accelerated in the early 1950s; however, major problems still existed. From 1944 through 1957, the Arkansas Agricultural Experiment Station and the fibers section of the Agricultural Marketing Service of the U.S. Department of Agriculture tried to deal with the problem that machine-picked cotton graded one and a half grades below hand-picked. Scientists grappled with the problems of leaf stains, excess moisture, trash and other foreign materials, twisting and tangling of the lint and discoloring from spindle oil.

The changing face of Arkansas agriculture was monitored by rural sociologists William H. Metzler and J.L. Charlton. In the Arkansas Ozarks, only 18% of the families were still totally dependent on agriculture by 1956. Row crop vegetables had disappeared in favor of poultry, fruits, livestock and dairying, which allowed part-time, off-farm employment.

Agricultural scientists believed farmers had failed to develop the agronomic potential of their unimproved land. Planting pine trees on submarginal lands allowed farmers to derive some income from wood lots. Lack of information, poor cutting practices and lack of fire control hampered the development of economically sustainable wood lot management. Many farmers allowed livestock to forage in the woods, destroying valuable seedlings. The process of long-term wood lot management with planting of commercially valuable species could give farmers a supplemental source of income.

The Hatch Act of 1955, which consolidated the Hatch Act and supplementary legislation, had renewed the traditional emphasis on basic research. Congress both simplified the accounting system of the Stations and committed itself to higher levels of funding. As a consequence of additional money, the Experiment Station in 1958 built a physiology and processing laboratory to examine the effects of pre- and post-harvest variables on the quality of processed vegetable and fruit products. Programs included soybean utilization and soy oil biochemistry, viticulture, food microbiology, rice utilization, cereal chemistry, vegetable acidification and pickle biochemistry.

The poultry industry, with its integrated production techniques, needed sophisticated studies of capital investment due to the low profit margins. The search for cheap sources of feed led E.L. Stephenson to test hydrolyzed poultry feathers and poultry by-products as a ration supplement.

The demand for improved varieties led the Experiment Station to engage in foundation seed production and to build a seed processing plant at the Rice Branch Station in Stuttgart. The plant was designed by Arkansas Experiment Station engineers to insure that seed growers would receive the highest possible quality of foundation seed of recommended varieties of rice, soybeans and small grains. The varieties developed included 'Kamrose' and 'Prelude' rice and 'Taggart' and 'Ferguson 560' oats. The Experiment Station worked with the State Plant Board and growers to produce and sell disease-free strawberries.

John Bowers and M.J. Goode worked on disease-resistant varieties of cucumbers in the region around Atkins, Arkansas. Victor Watts developed a new watermelon variety, the 'Hope Diamond', which was resistant to fungus and leaf spotting disease. The drought cycle brought an interest in supplemental irrigation of fruits and vegetables. A.A. Kattan demonstrated that, by maintaining soil moisture above 50% and spacing tomato plants 18 in. apart, farmers could achieve a significant increase in yields. In 1955, Experiment Station scientists, with some assistance from local tomato growers,

M. Isabel Irwin (shown here supervising Wayne Hamm in a food lab) found that linoleic acids found in corn oil could prevent hardening of the arteries. (Photo courtesy Photograph Collection, Special Collections, University of Arkansas Libraries.)

started a series of experiments to evaluate new varieties of tomatoes that would not be susceptible to Fusarium wilt and cracking.

In 1958, home economist M. Isabel Irwin initiated some of the first experiments dealing with the role of lipid cholesterol, lecithin and triglycerides. Preliminary evidence derived from her work indicated that linoleic acids found in corn oil could play a part in preventing hardening of the arteries.

The growing sophistication in soil analysis led to a full-field testing center on the Cotton Branch Station. The field, named the Nelson field after Martin Nelson, would be the gauge to see if chemical application and crop rotation changed soil quality.

By 1958, soybeans were the second most valuable crop in the state, generating almost $100 million in Arkansas. The release of 'Lee' in 1953 in cooperation with the U.S. Department of Agriculture Southern Regional Soybean Laboratory provided farmers with a soybean variety that did not shatter at maturity and made soybean growing a profitable enterprise in the South. This valuable crop was increasingly endangered by the appearance of soybean diseases such as the soybean mosaic, bud blight, bean yellow stipple and bean pod

The area known as Agri Park was formerly a part of the Station's dairy pasture but was separated from the main pasture when Highway 112 was relocated in 1948. Lippert S. Ellis (shown above) stands in front of Barton Pavilion in Agri Park, which was named in honor of Col. T. H. Barton, chairman of the board of the Lion Oil Company in El Dorado, Arkansas, in 1957. A piece of lignum vitae beam from the Republic of Panama, engraved with that country's Coat of Arms, was built into the wall near the fireplace. Below the beam are six stones from the Old Panama City that was sacked by the pirate Morgan in 1670. The gift was made in appreciation of the College of Agriculture's assistance in the development of agricultural programs in teaching and research in Panama in the 1950s.

mottle. In 1959, the Experiment Station released a new variety of soybean named 'Hill', jointly developed with the U.S. Department of Agriculture Southern Regional Soybean Laboratory, which was highly resistant to many common diseases and matured early to allow farmers to have a second crop of small grains.

In response to the expansion of the cattle industry, the Experiment Station leased land to explore the problem of maintaining production of permanent pasture forage during the hot, dry summer months. The sites were used to test varieties of grasses to ascertain their nutritional efficacy and hardiness. Production of alfalfa increased for hay from 35,000 acres in 1952 to 60,000 acres in 1955. L.H. Hileman, R.L. Beacher and C.E. Wells tested for the response to trace minerals, boron, copper, zinc and molybdenum as well as phosphate, potash and lime.

Associate Director John White felt that the challenge of agricultural research in Arkansas was complicated by the major shifts in traditional cultivation. The stations in eastern Arkansas concentrated their efforts on cultivating, mulching, drainage and tillage practices in the heavy, moist soils of the Delta, which call for special agricultural techniques. In 1957, the 634-acre Fruit and Truck Branch Station at Hope was refurbished to include new projects involving soybeans, rice, oats, wheat, corn and grain sorghum. In addition, tests were conducted in forestry and pasture management.

With the Congressional mandate to assist farm marketing, a cooperative market news service opened in Little Rock as well as a statewide report from the livestock market in Fort Smith. Even with these new sources of news, John White believed that demands being made on the Station were too great compared to its limited resources. Nevertheless, Arkansas farmers looked to the Station for scientific information to help them solve their production, handling and marketing problems.

The synergistic nature of agricultural research could be seen in the work of Maurice Ray in animal husbandry who fed steers with poultry litter and production byproducts, which proved an effective replacement for cottonseed meal. Grain sorghum rapidly increased as a result of soil diversion programs and provided an alternative source of cattle forage. Joseph Keogh and Richard Maples commenced experiments to test nitrogen fertilizers.

In 1959, Lippert Ellis was replaced as Station director by E.M. Cralley, a specialist in rice disease who had served 28 years on the Station staff. Cralley emphasized the need for cooperative research with the U.S. Department of Agriculture to develop multi-disciplinary programs. Moreover, multidisciplinary research was necessary to make major advances. For example, in the field of beef production, agronomists studied grasses, legumes and fertilizers; entomologists worked on ticks, grubs and biting flies; agricultural economists

E. M. Cralley, an authority on rice diseases and culture and head of the department of plant pathology, succeeded Lippert Ellis as Station director in 1959 and served until 1973. He advocated multi-disciplinary research and cooperation with the U.S. Department of Agriculture, private industry and business.

examined profitability of production techniques and marketing; and the animal husbandry department worked on improved breeding and management.

To accomplish these goals, Cralley reorganized the Experiment Station into nine major departments, six branch stations and five substations. The Northeast Branch Station, on 627 acres at Keiser, had been established in 1957 through the expansion of the alfalfa substation that was located in Osceola. James Jacks was the first assistant director. Early research on the station included all phases of cotton and soybean production, alfalfa production, evaluation of experimental lines and varieties of corn, cotton, grain sorghum, small grains and soybeans.

The Southeast Branch Station, also established in 1957, occupied 634 acres near Rohwer. Kenneth B. Smith was named assistant director. The four major areas of research established at the station were cotton (irrigation, control of insects, management and culture), rice (varieties, grass and weed and control and fertilizer studies), soybean and other crops.

The Experiment Station worked closely with agricultural business to engage in product development by major commodity dealers such as Archer-Daniels, Midland and General Foods. Cralley emphasized

John W. White was the first native Arkansan and University of Arkansas graduate to be named the Station's associate director. He also served as superintendent of the Branch Station at Stuttgart and head of the department of agricultural economics and rural sociology. In 1959 he became the University's first vice president for agriculture and held that position until 1974.

that Arkansas farmers had to take full advantage of their niche in the agricultural economic system.

In 1960, the Experiment Station was making major contributions to agricultural progress. H.R. Rosen developed 'Ace', a new, soft winter wheat, with the assistance of R.L. Thurman and G.E. Templeton. The new wheat would be used as a forage for winter grazing because it was resistant to Hessian fly.

To improve the quality of rural life in 1960, the Arkansas Agricultural Experiment Station Water Testing Laboratory was organized to perform mineral analysis for rural households to improve quality of water. The lab analyzed the mineral content of well, lake and spring water.

Yet the triumph of the scientific revolution in agriculture had not reduced persistent surpluses nor prevented marginal farmers from leaving farming. The decade of the 1970s would present unprecedented challenges to the Experiment Station in a rapidly changing world.

The Station realized the importance of water quality in rural areas and established a water testing laboratory in 1960. Agronomist Les Hileman (standing) and other researchers tested water for irrigation, poultry and livestock and analyzed the mineral content of domestic water supplies from wells, lakes and springs.

Chapter VI

The Changing Face of Arkansas Agriculture in the 1960s

The decade of the 1960s was marked by major challenges to individual researchers in their disciplines and farmers as a socioeconomic entity. The shrinking demographic proportion of farmers in the population began to weaken Congressional support for agricultural research.

The new age of farming required increased expertise as well as greater amounts of capital. After 1960, the mechanized farm had unaccustomed flexibility. The farmer could use machinery to grow soybeans or grains as well as to plant grass and forage for livestock. The sharecropper was rapidly becoming a figure from the past. There was no room for the cotton farmer with 5 to 10 acres of cotton.

In the national arena, amidst conflict for Congressional funding, the National Academy of Science issued the Pound Report claiming that the Experiment Stations allowed too much duplication of effort and neglected basic scientific inquiry in favor of problems of practical application. In Arkansas, Experiment Station scientists tried to keep pace with the new and shifting demands made on them. The markets for Arkansas' agricultural products were to move from regional to national and ultimately to international levels.

The Experiment Station's research both shaped and was reshaped by external events. Arkansas corn acreage was reduced from 670,000 harvested acres in 1956 to 238,000 in 1961 due to the presence of the southwestern and European corn borers and the effects of federal grain programs removing land from cultivation. However, as a consequence of better seed varieties, fertilizers and cultivation practices, yields increased almost 20% in the same time period.

The process of research was time consuming. For instance, agronomist J.O. York and entomologist W.H. Whitcomb started preliminary studies in 1956 to deal with the southwestern corn borer. A program was started on the Vegetable Substation near Van Buren to find a variety of resistant corn. By 1962, the Arkansas Experi-

The Arkansas Station was one of the first state experiment stations to conduct beef bull performance testing in the 1950s, but tests weren't open to private cattlemen until 1962. Bull testing facilities are now located at Savoy (pictured above) near Fayetteville and at the research and extension centers in Hope and Monticello.

ment Station developed synthetic varieties from germplasm reservoirs. In 1965, tassel evaluation showed a high percentage of borer-free plants. Finally by 1966, the agronomists and entomologists developed a corn variety with a high degree of resistance to both the corn borer and the dwarf mosaic.

Funding was expanded for the branch station system. The Southwest Branch Experiment Station was formed in 1960 from the Fruit and Truck Branch Station at Hope. The station was originally created in 1925 in Hope when 30 businesses and 44 individuals contributed $87,000 to purchase 185 acres. In 1961, the Experiment Station leased the 1,000-acre "Spencer Tract," which it later purchased. In 1962, Connell Brown initiated the Beef Bull Performance Test in which private owners nominate and place beef bulls at the test facility for determination of growth data, average daily gain, feed efficiency and weight per day of age.

In 1981, the Southwest Branch Experiment Station became the Southwest Research and Extension Center with continuing studies in cow herd management and systems of feeder calf production; effects of the conversion of pine and hardwood natural stands to pine plantations as well as the management of existing natural pine and hardwood stands; small grain, corn and grain sorghum varieties;

The Alfalfa Substation, established in 1948, became the Northeast Branch Station near Keiser in 1957. That same year the Southeast Branch Station was founded near Kelso. In the late 1970s, the Arkansas General Assembly appropriated funds to establish Research and Extension Centers at Hope, Keiser (above) and Monticello.

cover crop production; cropping systems; and soil testing. Horticultural projects have included watermelon, southern field peas and breeding and selection of pickling cucumber varieties.

In the 1960s, new crops continued to supplement old favorites. For instance, the pioneer work of H.R. Rosen in wheat adaptability led to a doubling of wheat acreage. In soybeans, state production increased from 12 million bu in 1950 to 87 million bu in 1969. The processing of beans became a major industry in Mississippi and Phillips counties. On the other hand, cotton producers faced a combination of land retirements, a squeeze on profitability, growing competition from the southwest and substitution of synthetic fibers for natural fibers, all of which contributed to the decline of cotton as the major cash crop in Arkansas. After a study on the future of cotton, B.A. Waddle concluded that even with recommended varieties, irrigation, herbicides, fertilizers and improved spacing, there existed a natural ceiling for cotton production in Arkansas. A new cotton variety was released; 'Arkansas 61-28' had a short growing season, was storm resistant and could be harvested with a mechanical picker.

Another major crop in Arkansas, rice, had experienced considerable growth due to opening of export markets in southeast Asia. The Rice Station at Stuttgart was joined by the Southeast Branch and Northeast Branch in cooperative participation in rice research to study the entire range of rice cultivation, including breeding, weeds, insect and disease control, fertilizer applications and the marketing of rice and rice products. Two new rice varieties were released in 1963: 'Nova', a medium-grain variety, and 'Vegold', a long-grain variety. Both had a relatively short growing season; however, the new varieties needed nitrogen top dressing for maximum yields. This discovery was coincident with the work conducted by Vernon Hall and Dewey Tackett of the Rice Experiment Station on the nutrition balance provided by nitrogen.

As a major research arm of the state and cooperating with the Federal Government and the state, the Experiment Station expanded the umbrella of its research endeavors. For instance, the potential of rural Arkansas for recreation was probed by Max Jordan (Bulletin 673). Tourism was beginning to become a major business with the building of the interstate highway system and the creation of large artificial lakes by the damming of the Arkansas and White Rivers. Nevertheless, Arkansas had major obstacles to overcome. Lack of capital, poor advertising of attractions and lack of adequate liability insurance all inhibited growth. Nevertheless, with continued poor prospects in agriculture, the economic investments in lakeside motels, fishing ponds and restaurants presaged a future economic alternative for the traditionally poverty-stricken uplands.

The 1960s continued the accelerated mechanization of agriculture, the displacement of tenants and a movement to economies of scale with the average farmstead increasing in size from 50 acres to 300 acres. For example, the cotton growers in eastern Arkansas needed use of an ammonia applicator, ammonia tank, cotton thinner, cotton trailer, crop duster, cultivator, disk harrow, plow, grain drill, lime and fertilizer spreader, middle buster, mold board plow, planter, picker, cotton rotary hoe, spike tooth, spring tooth harrow and spray rig as well as a tractor. The intensive agricultural practices in California and other states kept all but the most efficient Arkansas farmer in a profit squeeze. Moreover, the cycle of periodic droughts placed an emphasis on irrigation in order to protect yields from the stress of the periodic lack of rain.

The decline of rural America in the face of urban growth resulted in an awareness of the need for land-use planning. Arkansas, respectful of the inherent rights of private property, shied away from compulsory dictates of the government. Although county planning boards had been created in 1937 in response to New Deal legislation, they remained strictly advisory. The Experiment Station, working in conjunction with the University of Arkansas Law School, studied the alternative land usage pattern with emphasis on recreational and forest conversion in order to attain an improved quality of life.

The new Kennedy administration with Orville Freeman as Secretary of Agriculture had promised new direction in agricultural policy. Soil banks with their land retirement provision and flexible price supports had not brought prosperity to the small farmer nor stemmed the rural exodus. In Arkansas, severe limitations on rice allotments accelerated the shift to soybeans. The return of the periodic droughts of the 1950s had convinced the more affluent farmers of the efficacy of irrigation.

The beef industry expanded, especially in northwest and central Arkansas, but the limitation on energy feeds resulted in the shipping of 70% of all cattle out of the state for finishing. Pasture management as well as improved breeds of cattle with high feed efficiency conversion was important for an economically competitive beef industry. David Womack helped develop a new hybrid of silage sorghum with a high percentage of protein. Agricultural economist Lloyd Bender concluded that if farmers could not either produce more corn or increase feed efficiency, they would be at a marked disadvantage compared with other producing areas. Similar studies by Lee Martin confirmed that Ozark cattle producers would not have sufficient levels of profitability unless the mortality and feed conversion rates were lowered and quality improved.

Connell J. Brown in animal sciences continued ongoing studies to reveal sources of variation in the weight of beef calves such as the season of birth, sex and age of the dam. M.S. Offutt examined the problems in conversion of land to pasture. Agricultural scientists worked on determination of the protein content of forage and studied the effect of slaughter techniques on meat quality. Maurice Ray in animal industry predicted that, with improved forage quality, Arkansas could produce beef that could provide for in-state supermarkets.

To achieve this goal, Erwin Auxiliary Airport in Newport, formerly used to train World War II pilots, was acquired in 1959 as the Beef Substation to test pasture management and management of beef calves after weaning. In 1961, the Experiment Station started central testing of cattle to assist breeders; the Main Station had expanded its own long-term breeding program and was participating in regional beef cattle breeding studies. In addition, the researchers at Hope began studies of several species of forage plants to establish base lines for pasture improvement.

Accompanying the increased cattle industry was increased interest in alfalfa as forage for dairy use. As early as 1895, the Experi-

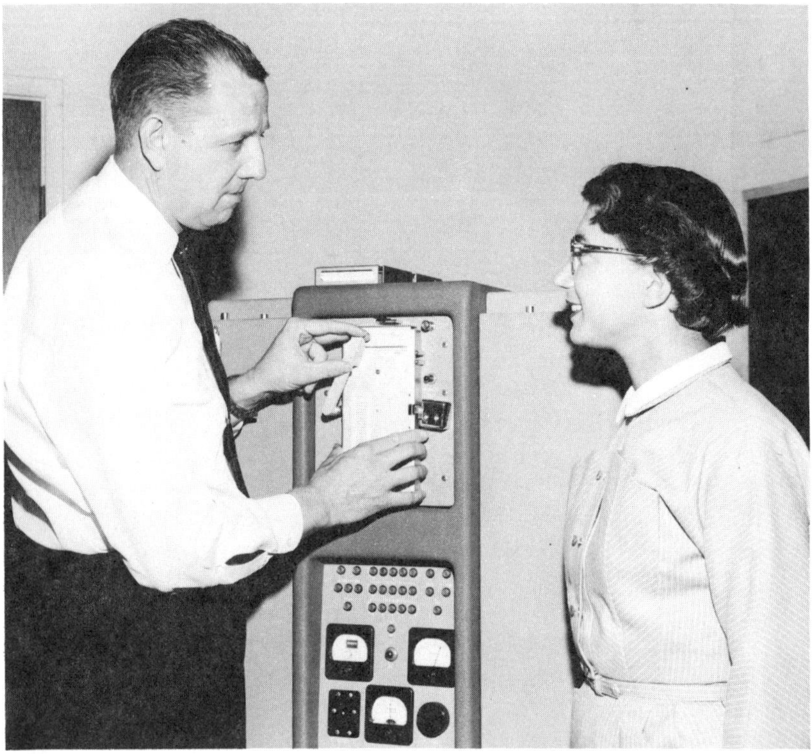

Animal scientist C.J. Brown obtained the first digital computer in Arkansas in 1957 to compile statistics on the influence of heredity on beef cattle weight and body characteristics. Brown is pictured here (on left) receiving training in the operation of the new computer from an instructor at the Bendix Corporation in Santa Monica, California. (Photo courtesy C.J. Brown.)

ment Station had recommended that farmers grow alfalfa. In 1897, G.L. Teller had started fertilizer experiments to test the efficiency of nitrogen, phosphate and potash on the plants. R.L. Bennett had been a fervent advocate of the raising of alfalfa, and Martin Nelson in 1929 had incorporated alfalfa in his crop rotation tests. By 1958, L.H. Hileman was studying the need of alfalfa for micronutrients such as boron. Conversion of eroded old cotton land to pasture offered a potentially more profitable alternative to row crops.

To assist cattle producers, experiments were conducted to create year-round pastures, which included Bermudagrass, fescue, white clover, Bermudagrass-vetch and Bermudagrass-rye grass combinations. Also the Experiment Station tested Sudan-sorghum hybrids

The Beef Substation near Newport, formerly an air field for training pilots during World War II, was established in 1959 to provide needed acreage for studies in pasture management. It also has facilities for the performance testing of bulls and boars. (Photo courtesy C.J. Borwn.)

for forage. By 1967, bermudagrass had become the major summer base grass in the state.

The Southwest Branch Station had also become involved with cattle experiments to ascertain methods for increasing carrying capacity of pastures. With the dramatic growth in feeder calf production, forage and feed experiments became increasingly important. Maurice Ray, A.E. Spooner and R.W. Parham showed that the carrying capacity of pasture could be expanded. Coordinated with these activities, long-term research was initiated on Hereford and Angus cattle lifetime growth patterns. Spooner and Herbert Huneycutt examined the possibility of using Johnsongrass for pasture.

The traditional agricultural way of life was changing, accompanied by an increasingly negative attitude regarding rural life and farming. Many Arkansans despaired of making a living solely by farming. Moreover, a skeptical view towards scientific farming remained in the more remote parts of Arkansas. A survey of a farm community in Van Buren county by rural sociologists indicated that only 38% of the farmers thought that a college education was necessary for farming, and 8% doubted if even high school was necessary (Bulletin 650).

The publication of Rachel Carson's book, *The Silent Spring,* in 1959 made people aware of possible ecological damage by indiscriminant use of DDT and other pesticides. Arkansas Experiment Station entomologists had long since adopted a cautious view of the application of pesticides. As early as 1922, Dwight Isely had warned against upsetting the balance of nature. To develop a sensitivity towards the timing of application of insecticides, Isely had started the practice of cotton scouting to study insects involved, recognize their potential damage and evaluate methods of control. By the early 1930s, county extension agents were training farmers to do their own scouting. In 1942 Charles Lincoln had started scouting schools for farmers and trained the first full-time scouts. By 1960, county agents had 42 scouting schools training over 8,000 farmers in 25 counties. The movement to have professionally trained scouts started in 1949, and by 1961 some 130 scouts working about 10% of the cotton crop augmented farmers who had been trained by the Extension Service. As a consequence of this effort plus improved varieties, fertilizer and cultivation practices, lint production per acre had increased from 200 lb/acre in the 1920s to 515 lb/acre by 1960.

The Kennedy administration saw food exports, especially to Latin America and Asia, as an important ingredient in foreign policy. The Food for Peace Program and the Alliance for Progress both envisaged putting America's surplus to work fostering good will in the Third World. Not only staples, but also vitamin and protein enrichment of basic export commodities were stressed. Leslie Easterling saw fish supplement in rice as a way of combatting protein deficiencies, and the soybean, an excellent source of protein, was to benefit from the enlargement of America's overseas trade.

Soybean production continued to expand. Prices were maintained at favorable levels relative to other crops, and soybeans could be easily harvested mechanically. After 1950, cotton oil mills were modified for production of soybean oil. A storage problem was caused by the relatively short harvest time made possible by combines, but the problem was solved by the proliferation of new facilities. E.M. Cralley reported that the Experiment Station was in the forefront of soybean studies. Every aspect of soybeans--enzyme system, herbicides, diseases, resistant varieties, weed control, irrigation, seed bed preparation, depth of plowing, row width, spacing and fertilization--was under study by agronomists and scientists in other disciplines. Plant pathologists were interested in virus diseases, seedling diseases, stem canker and nematodes. Xzin McNeal in agricultural engineering developed a method of storing soybeans with reduced moisture and a temperature below 60°F in order to prevent germination. Floyd Miner coordinated methods for the control of soybean stink bug by both biological control and chemicals.

The continued shift to soybeans resulted in the displacement of cotton as Arkansas' leading crop. The Experiment Station released 'Lee 68', a soybean variety highly resistant to phytophthora rot, as well as 'Davis'. Experiment Station scientists continued their soybean research efforts, especially in view of projections of future global demand for this highly adaptable plant.

While new programs were being initiated, the decade of the 1960s marked the termination of crop rotation and fertilization experiments that had been started in 1934 under Martin Nelson to test nutrition depletion by corn, hay, wheat and oats. By 1941, adapted hybrid corn had replaced open pollinated Paymaster varieties. After 1948, lespedeza was seeded over the entire tract. The carefully examined data gave baseline information on optimal amounts of phosphate, lime, nitrogen and potassium to be utilized by farmers.

Following the pest control strategies of Dwight Isely and Charles Lincoln, entomologist J.R. Phillips expanded the concept and developed community-wide integrated pest management programs in cotton.

With farmers no longer solely dependent on cotton, Experiment Station scientists turned to fruits and vegetables. Agricultural economist Carter Price suggested that irrigated truck farming was the most efficient method of generating substantial income given that 51% of Arkansas commercial farmers had less than 50 acres of cropland (Bulletin 732). In coordination with the Arkansas Industrial Development Commission and Arkansas Power and Light, seven farmers participated in experiments raising pepper, squash, okra, strawberries and tomatoes. Roy Rom and Wayne Sabbe in 1968 announced a new program of leaf analysis for fruit growers to assist them in reaching higher levels of efficiency through better fertilizers. Cross breeding of tomatoes and peppers was undertaken. In 1962, the Station released a new spinach variety, 'Savoy Hybrid 612', in coordination with the Crops Research Division of the Agricultural Research Service and the Texas Agricultural Experiment Station. Spinach was to become a major crop in the Arkansas River valley.

As Arkansas revived its truck farming efforts, competition from other sections of the country hampered the sales of Arkansas exports. Agricultural economist Carter Price commenced a series of studies to discern the reasons for prejudice against Arkansas peaches and tomatoes. Applied marketing information provided valuable information for plant scientists.

Horticulturist Roy Rom (left) and agronomist Wayne Sabbe in 1968 developed a new program of leaf analysis for fruit growers. Rom also conducted research to determine which varieties of apples would be best adapted to Arkansas, and in 1984 he helped to develop two new varieties of peaches, 'Allgold' and 'Goldilocks', for the processing industry.

A major impediment to the development of a berry industry in Arkansas was the lack of availability of cheap, reliable hand labor necessary for picking. As a result, the Experiment Station agricultural engineers and horticulturists commenced work on a mechanized blackberry picker. By 1965, agricultural engineers had developed a prototype blackberry harvester. Concomitantly, G.A. Bradley of the department of horticulture started to examine the possibility of commercial production of blackberries.

The effort to resurrect the fruit industry in Arkansas continued at the Fruit Substation at Clarksville. Originally, the fruit station had concentrated on peaches, but by the mid-1960s it had expanded to include apples, apricots, blueberries, cherries, grapes, nectarines, pears and plums. Dr. Roy Rom worked on an apple research program to determine which varieties would be suitable for Arkansas soil conditions and climate. In 1965, the personnel had started extensive irrigation studies. By 1967, the mechanical harvester for blackberries had been adapted for strawberries, which were also limited by the availability of labor. As a consequence, agricultural breeders at the Experiment Station started breeding strawberries more adaptable to mechanical harvesting.

In the late 1960s, agricultural engineers and horticulturists developed the country's first mechanical harvesters for blackberries and strawberries ((the strawberry harvester is pictured here).

At the Southwest Branch Experiment Station in Hope, L.H. Hileman, R.L. Beacher and Lyell Thompson started testing fertilization of fields with emphasis on soil acidity, nitrogen, potassium and calcium levels. Increasingly sophisticated instruments for chemical analysis allowed agronomists to broaden their focus in plant nutrition to include micronutrients such as boron, manganese, zinc, copper and iron. To more efficiently take soil samples, D.A. Brown and Billy Bryan developed an undisturbed soil core sampler that was both portable and battery operated.

Continuous farming in eastern Arkansas with irrigation had resulted in changes in soil chemistry and increasingly saline or alkaline soil. M.E. Horn warned that the process of possible crop danger was accelerating. Vernon Hall and Lyell Thompson found that rice areas were characterized by poor internal drainage. Similarly, lime was needed to combat high acidity in the soils of eastern Arkansas.

In the late 1960s, Joseph Keogh and Richard Maples commenced work on boron deficiencies, following up on initial observations of Lyell Thompson and Wayne Sabbe. Scientists reported that boron applications resulted in increases of seed cotton yields per acre. The scientists also tested calcium and organic matter as determinants of yield. To ease the difficulties of soil analysis, Keogh and Maples developed a fast-drying net hood that did not adversely affect soil tests. The technique of using forced air drying at 30°C in a soil box for 3 to 4 days allowed for easy analysis. By 1977, the team of Richard Maples, Joseph Keogh and Wayne Sabbe started to observe that over application of fertilizer resulted in the formation of nitrate salts that slowed seedling cotton and caused significant stand losses.

The Agricultural Experiment Station had a dual thrust on disease control. On one hand, scientists worked closely with the Cooperative Extension Service and with the State Plant Board to deal with the plant certification program. On the other hand, the Station concentrated on basic research to investigate why plants were injured by disease as well as soil-borne nematodes.

To aid the commercial timber companies, the life cycle of the Nantucket pine-tip moth, which attacked short-leaf and loblolly pines, was studied by Lloyd Warren. Entomologist W.P. Boyer predicted that new pests--the face fly, the southern pine beetle and the Japanese beetle--would probably be attacking Arkansas crops and livestock sometime in the future.

The purview of research was continually expanded. The Cotton Branch Station, established in 1926, had expanded to 895 acres by

1964. 225 acres were leased for a project on systems of finishing animals on feed crops adapted to the delta section of Arkansas. The objective of the study was to show that cattle could be grown on Arkansas grains. The diversification of Arkansas crops reflected itself in research emphasis on soybeans, small grains, corn and beef cattle. In addition, the Station had more than 2,000 different genetic cotton strains out of which 'Rex' and 'Rex Smoothleaf' cotton were developed.

A major research objective was creating inexpensive forage out of agricultural byproducts. Soybeans, mill-run rice hulls and chicken litter were tested as feed to help replace part of the 2.2 million tons of feed grain purchased by Arkansas farmers each year. Maurice Ray stated that rations containing broiler litter could be used suc-

VARIETIES AND STRAINS DEVELOPED BY THE ARKANSAS AGRICULTURAL EXPERIMENT STATION IN THE 1960s

AGRONOMIC CROPS

Barley	
Zora	1965
Corn	
AKC 66	1966
Cotton	
Rex Smoothleaf	1963
Rex Smoothleaf 66	1966
Arkansas 61-28	1968
Forage Sorghum	
Leafmaster 43	1963
Grain Sorghum	
AKS 614	1962
AKS 663	1966
AKS 653	1966
Lespedeza	
Summitt	1962
Oats	
Ora	1964
Nora	1966
Rice	
Northrose	1962
Vegold	1963
Nova	1963
Nova 66	1966
Starbonnet	1967
Soybean	
Davis	1966
Lee 68	1968

HORTICULTURAL CROPS

Cucumber	
Southern Pickler	1961
Southern Pea	
Crimson	1965
Mekan	1967
Spinach	
Savoy Hybrid 612	1962
Tomato	
Bradley	1961
Pope	1966
Turnip	
Crawford (greens)	1965

cessfully for feeding cattle, but each batch of litter had to be analyzed chemically for the crude proteins, fiber, dry matter and nitrate. In addition, the feed would have to be supplemented by at least one quarter protein. P.R. Noland and C.A. Baugus tested the efficiency of utilizing heat-damaged rice in growing and finishing swine rations. Their tests indicated that half of the corn ration could be replaced with ground, heat-damaged rice without adversely affecting performance.

In 1966, N.R. Gyles, J.G. Shook, J.E. Valentine and L.D. Andrews worked on methods of feed conversion for turkeys (Bulletin 710). With the profit margin only 2 cents/lb, improvement in poultry feed conversion was essential for profitability. By the 1970s the studies of turkey rations had grown to include the problem of pyridoxine deficiencies. A team of poultry scientists worked out the use of dried supplements to deal with protein deficiencies. By 1977, the Experiment Station arrived at optimal protein levels for turkey producers (Bulletin 821).

The Experiment Station continued its role as a cooperator in many ongoing federal programs. The Ozark Regional Commission and Experiment Station personnel assessed housing needs of the Ozarks poor. Rural sociologist Lloyd Bender dealt with the new trend of immigration into the Ozarks of retirees, "Back to the Landers" and previous outmigrants. Economic opportunities arose from new retirement communities and recreation. The pristine beauty of the Ozarks with its man-made lakes contributed to an influx of tourism, which was rapidly becoming a major Arkansas industry.

As part of the national statistical base to ascertain future progress, Hugh Spurlock of the U.S. Department of Agriculture Economic Development Division working in conjunction with the Experiment Station surveyed Northwest Arkansas to measure the percentage of houses falling under the governmental definition of "substandard" (Bulletin 736). Later studies followed up the impact of Federal programs for ameliorating poor housing conditions. Mobile homes, despite their drawbacks, were seen as a short-term answer to inadequate housing.

In 1968, Karol Budd and J.L. Charlton started to examine the impact of the School District Act of 1948, which abolished districts with less than 350 persons of school age (Bulletin 731). Despite these reforms, the continued deficiencies of rural schools caused the authors to recommend further consolidation and additional state and local aid. The continuing poverty of the inhabitants of the

Station researchers worked with the Ozarks Regional Commission and the U.S. Department of Agriculture Economic Development Division to evaluate the housing needs of the poor in northwest Arkansas. Rural sociologists Lloyd Bender and J.L. Charlton (right) examined the impact of the movement of retirees into the Ozarks and the status of rural schools, respectively. (Photo courtesy J.L. Charlton.)

Ozarks indicated that low career expectations, combined with limited resources, discouraged participants from breaking out of the cycle of relative deprivation. The educational needs of youth would be the key to promote economic development.

The Arkansas Agricultural Experiment Station with its research capacities played a role in ascertaining the impact of anti-poverty programs applicable to rural areas. The Experiment Station projected price trends in cotton, beef, rice and dairy with in-depth analysis of expected factors of production. Of course, the price projections could not anticipate the 1973 war in the Middle East and the dramatic escalation of energy prices by OPEC that rippled through the world economy. However, economists did foresee the continued decline of farm ownership due to increases in costs of mechanization.

An increase in Experiment Station funding from private and state sources allowed for expansion of operations. The Rice Branch Station in Stuttgart had been enlarged from its original 160 acres to 915 acres and released another new rice variety, 'Northrose', an early, stiff-strawed medium grain. Rice was rotated with soybeans, lespedeza, grain sorghum and small grains to provide a balanced agriculture.

In 1963, rice farmers, in an attempt to utilize land diverted from rice production by the Government Price Support Program, initiated the production of farm-raised fish on a commercial basis. The nascent industry encountered consumer resistance, and fish proved difficult to market. The Experiment Station started programs promoting Buffalo and catfish as attractive products. Fish farming studies were also introduced at RREC as a means of control of mosquito larvae.

The Southeast Branch Station, established at Rohwer in 1957, had 634 acres by 1965. The branch station studied control of bollworm and boll weevil by insecticides and by natural resistance. In addition, tests were made of grain sorghum, rice, small grains, corn and vegetable crops.

The activities of the Station reflected the transition in Arkansas' economic, social and demographic profile. In 1940, Arkansas had been the most rural state in the union with only 20% of the population living in towns of more than 2,500. By 1965, 50% of the population dwelled in an urban environment while small farming commu-

In the mid 1960s, Calvin R. Berry (on right) was a member of the Dairy Marketing Advisory Committee that conducted a major study laying the groundwork for the formation of large-scale, regional dairy cooperatives in the central and southeastern part of the nation. Berry is pictured with Joe Murphey, marketing specialist with Associated Milk Producers, Inc. (Photo courtesy C.R. Berry.)

nities were disappearing. Overall, the state population had fallen 8.4%, reflecting the impact of agricultural mechanization, persistent low returns to agriculture and the lure of higher wages in other parts of nation. Arkansas' demographic losses reflected the continuation of deep root poverty unable to respond to new farming techniques due to insufficient financial and human capital.

The problems of farming were compounded by the choices presented by the Food and Agriculture Act of 1965, which offered farmers the right to decide on "the level of production" as opposed to the level of allotment. Agricultural economist James H. White developed profit-driven scenarios that illustrated the various alternative methods farmers had for deciding on cotton as opposed to corn, soybeans, oats, wheat, grain sorghum, market hogs and beef cattle.

In 1965 the United States Congress passed an amendment to the Hatch Act to study the impact of pesticides on the environment. The expertise of the Experiment Stations was to be utilized in creating a regional center in Florida to test the health hazard of pesticides. In addition, in order to avoid duplication, a committee of nine was to review regional research initiatives on a state-by-state basis with the directors. The following year, Congressional concerns mandated Experiment Station efforts to reduce farming costs, better allocate water resources, improve human nutrition, deal with air pollution and assist farmers in establishing cooperative marketing networks.

The Arkansas Experiment Station developed a comprehensive series of tests on effectiveness of herbicides to ensure their efficacy and safety. Ronald E. Talbert and Robert Frans developed new techniques of "determining crop selectivity, desirability of further development and early dissemination of information to interested weed scientists." The Arkansas program received extensive funding from private sources.

The Experiment Stations, as well as the entire agricultural sector, were given an additional burden with growing public concern for the ecological and carcinogenic qualities of popular pesticides and growing fears of the unanticipated consequences of the chemical revolution. The ban on DDT led to a heightened awareness of the need for natural controls. The search for "safe levels" led to investigation in mosquito control, timing insecticide applications and modification of plant susceptibility by fertilization and cultural practices. Entomologists sought to study pest life cycles, dispersal patterns and pest-predator activities as a natural management system. Agronomists tried to breed plants that would exhibit pest resistance.

Robert E. Frans, holder of the Richard S. Barnett, Jr. Chair for Weed Science, developed ways to test the effectiveness of herbicides and to determine their impact on the environment.

Director E.M. Cralley saw the Experiment Station's role expanded beyond just commercial farming to conducting studies on the entire panoply of rural concerns. Cralley pointed out the need for studies in water quality standards and mosquito control and emphasized the importance of ecological studies. These shifting priorities echoed Congressional shifts in funding formulas. As a result, somewhat smaller proportions of research money were allocated for crop and animal resources with an increase in natural resource, people, community and institutional money.

Beef performance testing became one of the most successful programs of the Station. C.J. Brown in 1969 concluded preliminary tests on the heritability of birth weight, gains and weights during suckling period, transition and post weaning. Brown analyzed feed consumption and conversion to ascertain the genetic aspects of growth rate and feed utilization of beef bulls. In 1973, these studies also incorporated Angus, Hereford, Charolais and Santa Gertrudis sired crossbreds steers produced by Angus and Hereford cows. Brown also conducted studies on the relationships among measures of growth in Hereford and Angus cattle (Bulletin 773) and the influence of mature weight and rate of maturing on individual beef cow efficiency (Bulletin 774). By 1975, Brown, O.T. Stallcup and Z. Johnson had worked out predictive equations for estimating the net energy values of forage.

Plant pathologist George Templeton (left) and U. S. Department of Agriculture weed scientist Roy Smith (right) discovered a fungus that attacked the weed northern jointvetch in rice and soybeans. After a dozen years of research and development, the fungus was developed, in cooperation with the Upjohn Company, as the first mycoherbicide commercially available to control a weed in an agronomic crop.

In 1969, Roy Smith, the U.S. Department of Agriculture scientist stationed at the Rice Branch Station, noticed that a disease attacked northern jointvetch, a weed presenting serious problems in rice and soybean culture. He and George Templeton worked for 13 years developing the mycoherbicide developed by Upjohn and produced by Ecogen Corporation under the name "Collego." Collego has become a model for research on development of biological weedkillers around the world. The Station has continued its investigations into mycoherbicides, giving farmers an environmentally acceptable technique for weed control.

Technology and chemistry helped propel the agricultural revolution of the post-war period. In 1969, Roy Rom advocated the use of pre-emergent herbicides to control weeds in orchards. Agricultural engineers continued work on a prototype mechanical strawberry harvester. Accompanying these machines was an attempt to concentrate ripening fruit in order to insure "once-over" harvesting. Collaterally, Glen Nelson and A.A. Kattan were working on mechanical grading of strawberries. In addition, Station scientists released a new tomato, the 'Traveler', which was wilt resistant.

In 1968 the horticultural food science department was organized around the staff of the physiology and processing laboratory. Members of the department were working on quality control in the processing and handling of fruits and vegetables to overcome the negative perceptions of Arkansas fruits and vegetables.

In agricultural engineering, new research objectives included evaluation of cotton beds, designing economical optimal environments for poultry, meat and egg production and designing silos for storage of grain sorghum. Xzin McNeal helped develop an inexpensive method of storing cotton on a pallet trailer that was to be widely adopted. Plant pathologists started the screening of cotton stocks for resistance to wilt, particularly by discovering natural resistance. They tried to determine the ecological factors of cotton diseases caused by fusarium wilt fungus.

By 1970 the agricultural revolution had transformed American as well as Arkansas agriculture. However, fewer than 80,000 farms still operated in Arkansas. The next decade would see agricultural research face new challenges that could scarcely be foreseen. The Experiment Station would have to not only continue its traditional lines of inquiry, but also confront the ongoing energy crisis and its ramifications.

Chapter VII

Advances and Controversies in the 1970s

By the 1970s, agricultural research had achieved unimagined success. The United States had the capacity not only to supply its domestic market with inexpensive food and fiber but also to serve as the prime exporter to the world. Nevertheless, the decline in the marginal family farm and the accelerated concentration in corporate farming caused critics to question the thrust of scientific farming. Moveover, public fears increased as to the long-term, adverse effect of pesticides and herbicides on the ecological balance of nature.

In Arkansas, the shape of agriculture was accommodating itself to the new economic and political environment. The dramatic rise in meat prices intensified interest in beef raising. Plant breeders released 'Victoria', a new, multipurpose alfalfa with high disease resistance able to survive harsh winters. J.L. Perkins developed a practical way of bringing large numbers of ewes into estrus. To help farmers deal with drought cycles, A.A. Kattan worked on a scientific model for optimal levels of soil moisture for horticultural crops. The Station cooperated with the Fish and Wildlife Service in studies in aquaculture in conjunction with the Rice Branch Station and became actively involved with this science on the establishment of the program at the University of Arkansas at Pine Bluff. Acreage devoted to catfish production in Arkansas increased from 1500 in 1965 to 10,000 in 1970. Scientists grappled with oxygen depletion, diseases and parasites, "off-flavor," undesirable species of fish, toxic chemical residue, harvesting and marketing.

One of the major changes in Arkansas agriculture that transformed the state was the expansion of soybean cultivation. By 1969, about 4.2 million acres, 60% of Arkansas' arable land, was planted in soybeans. The collapse of the fish industry in the waters off Peru eliminated fish meal as an alternative protein supplement for cattle and provided expanded market opportunities for soybeans. Moreover, with rice and cotton restriction programs, soybeans as an alter-

In 1971 Station soybean breeder C.E. Caviness released a new variety of soybeans named 'Mack' after C.K. McClelland, who developed the first soybean variety adapted to Arkansas in 1928. Caviness also developed several other soybean varieties, including 'Davis', 'Hood 75', 'Lee 68', 'Lee 74', 'Jeff', 'Narow' and 'Lloyd'.

native became a $200 million a year industry. The growth in the soybean industry led to the expansion of elevator processing mills by major producers such as Riceland Cooperative.

With soybean production becoming a leading sector of Arkansas agriculture, C.E. Caviness in 1971 released a new variety of soybean, 'Mack', named after C.K. McClelland. Director E.M. Cralley predicted that the new variety, which was resistant to soybean cyst nematode and rot, would add $18 million to the harvest.

In 1972, the Experiment Station joined the U.S. International Biological Control Program to forestall proliferation of major soybean pests. W.C. Yearian, J.M. Livingston and S.Y. Young started tests on reducing green clover worm and other pest populations by optimal application of insecticides.

By 1970, the continued cultivation of cotton resulted in a fall in the nitrogen level in the soil. The Experiment Station analyzed soils by indirect means by measuring the organic matter in the soil and estimating the amount of nitrogen that would be released through the decomposition of organic matter. Agronomists studied nitrogen rates to determine the optimal level of fertilization with different cotton varieties and different soil types.

New projects in horticulture and forestry were initiated in 1971. The Experiment Station started to evaluate turfgrasses, evergreens, shrubs and trees that could be adapted to the Ozarks. Disease studies commenced on threats to loblolly and short-leaf pine trees that were increasingly replacing scrub oaks in the Ozarks.

The forest research effort was bolstered by the efforts of Faye Meade and A.T. McDaniel to develop Scotch pine for use as Christmas trees. Other scientists who worked in forestry included Harvey Holt in herbicide control and Phillip Pope and G.L. Wheeler in nutrient cycling of pine forests. In 1977, J.B. Murphy started experiments in the physiology of forest tree seeds. One of the most promising areas was F.H. Huang's program in tissue culture and biotechnology.

The Rice Research and Extension Center expanded in 1974 with a new headquarters building to house the resident director, professional Experiment Station researchers, Cooperative Extension experts and U.S. Department of Agriculture researchers. New varieties of rice were developed: 'Bonnet 73' in 1973, 'Nova 76' in 1976 and 'Mars' in 1977. The new rice varieties were released with complete fertilization management packages. However, the expansion of rice culture to many different soil types and the increase in fertilizer usage resulted in problems of rice diseases such as sheath blight, kernel smut and rotten neck blast. B.R. Wells' research on the use of zinc in rice production made a valuable contribution to higher yields.

The interdisciplinary approach to rice production resulted in a series of "How to Grow" bulletins by Bobby Huey and the staff at the center. The bulletins included seeding rates and dates, weed control, nitrogen fertilizer rates and timing to receive the "optimum response" from the new varieties. The center, under the direction of Francis Williams, also continued experiments with soybeans, wheat and oats, developed a major library on rice research publications, built greenhouses and laboratories and installed an underground pipeline system to allow measured irrigation.

As early as 1969, Congress started to discuss the involvement of the Experiment Station in the changing patterns of technology and the impact of technology on rural residents. To assist Experiment Stations in directing their research efforts, an Advisory Committee on Rural Development was created to be funded through new appropriations. Following this initiative in 1972, Congress enacted the Rural Development Act to give the land-grant institutions an additional \$10 million for 1974, \$15 million for 1975 and \$20 million for 1976.

The decade of the 1970s witnessed an intensified awareness of the potential health threat of pesticides. Agricultural scientists sought alternatives to potentially harmful chemicals. Promising alternatives developed by insect physiologists and pathologists included the use of sex attractants and viruses for control of certain pine, cotton, soybean and vegetable insects; strip-cropping to enhance the population of desirable predators; flame cultivation for control of weeds; and early varieties of cotton to shorten time of exposure to production hazards.

State support allowed for upgrading of facilities. In 1971, the Experiment Station added new research facilities in horticulture with an analytical laboratory for maturing and quality evaluation. The purpose of the new facility was the breeding of cucumbers, okra, southern peas, brambles, tomatoes, grapes, strawberries, apples, peaches and ornamentals and evaluation of the new crop potential of fruits and vegetables.

Cooperative Bull Testing continued to make a major contribution to genetic improvement in cattle as well as selective breeding of cattle that could efficiently convert feed into poundage. The Arkansas cattle industry ranked twenty-first in the nation as most farmers viewed beef production as a method to augment other farm operations such as poultry or as a method of supplementing income derived from other sources. Hereford and Angus herds maintained by the Experiment Station provided guidelines for the daily energy costs for growing and maintaining beef. The Herefords and Angus were raised as a "cow-calf" operation with the calves being sold off to feed lots. In addition, crossbreeding of Charolais and Santa Gertrudis bulls with Angus and Hereford resulted in steers that were faster growing with less fat and more muscle.

Portions of the Ozarks were denuded of brush and trees for use as pasture. The Experiment Station assisted farmers in selecting the grasses that would provide the necessary nutrients for cattle as well as survive the vagaries of Arkansas weather. Long-term studies on winter legume trials in Hope and Fayetteville resulted in recommendations of winter legumes for forage, hay, winter cover and green manure. In 1972, M.S. Offutt and L.H. Hileman studied bromegrass as forage due its high quality and superior drought tolerance (Bulletin 776). The Experiment Station estimated that Arkansas farmers experienced a 50% increase in forage production per acre over 1966 figures. The combination of cheap fertilizers from chicken litter with fescue grass and white clover as pasture proved an excellent financial opportunity for farmers. In 1974, A.E. Spooner and Maurice L. Ray recommended that a mixture of fescue and bermudagrass, fertilized four times in a year, could provide an excellent forage (Bulletin 791). The Experiment Station recommended bermudagrass interspersed with fescue to slow erosion on the rugged hillsides of Northwest Arkansas.

In the late 1970s, the Livestock and Forestry Branch Station under Dr. Kenneth Harrison expanded its activities. The station, which was established in 1937, explored the problems of livestock production in the Ozarks. It consisted of two eroded tracts, one of 2,062 acres on cotton land and another 887-acre forestry tract. Arkansas had been involved in projects to check erosion and create permanent pastures and meadows as part of the Bankhead-Jones Act. The Soil Conservation Service laid out a 40-acre tract for a drainage system to deal with periodic flooding.

In 1941, 40 purebred Aberdeen Angus heifers were purchased by the station as part of the first federally approved project on beef cattle performance. In 1964, most of the beef cattle were transferred to the Land Use Project at Colt, Arkansas, which was renamed the Pine Tree Experiment Station. The Livestock and Forestry Branch Station concentrated on dairy management problems with Jersey and Holstein dairy cows. The cattle were part of the Cooperative Regional Breeding Project in which artificial insemination using "top bulls" was compared with natural service with selected bulls. In 1977, a project was undertaken to chart the effects of breeding upon milk composition by artificial insemination. In 1980, the Holsteins were sold, and the Jersey herd was reduced to 50 cows.

In 1977, the Experiment Station organized the Southeast Research and Extension Center on the campus of the University of

Most of the Station's beef cattle in 1964 were transferred to the Land Use Project, now called the Pine Tree Station, near Colt. It has the largest land area, 11,880 acres, of all the Station's research units. Besides beef cattle, research is conducted on soybeans, rice, wheat, feed grains and upland timber.

The Southeast Research and Extension Center in Monticello was established in 1977 to strengthen research and extension programs of the region. It also has facilities for the Station's beef bull performance testing program. (Photo courtesy C.J. Brown.)

Arkansas at Monticello. The center was planned in the wake of a merger of Arkansas A and M into the University of Arkansas system in 1971. State Senator John F. Gibson of Dermott and James Scott, a former member of the Board of Trustees of Arkansas A and M, helped garner state funding for beef bull testing facilities and a new agriculture building at the University of Arkansas at Monticello. In 1977 the Experiment Station took formal control of the bull test facilities.

In January 1978, Gerald Brown, formerly of R.J. Reynolds Tobacco Company, was named director of the Southeast Research and Extension Center. Hugh Crowley from Auburn was hired to conduct research in soybeans, cotton and pastures. J.A. "Pete" Hornsby was given supervision of the beef bull performance test and a genetic by environment interaction study initiated with beef cattle. Terry Keisling was employed in 1979 to improve soybean cultural practices, and Walter Slabaugh studied soybean diseases. In 1982, Allen Mitchell was hired to examine questions relating to soil and water management for cotton production.

The effect of increased farm production rippled through the entire state's economy. The purchase of petrol and chemicals for agriculture was a half billion dollar business. In Arkansas, 62 agricultural cooperatives assisted farmers in buying and marketing agricultural

A hatchery building and ponds used in fisheries research at the University of Arkansas-Pine Bluff are in the background, with soybean and sunflower research plots in the foreground. Station researchers were first to promote aquaculture in Arkansas. Today research with catfish and other farm-raised fish is conducted at the Pine Bluff Research Station.

commodities. Processing plants, especially in poultry, became major employers.

In 1972, the University of Arkansas at Pine Bluff started major programs of agricultural research. The traditionally underfunded land-grant institution started to receive sufficient money to undertake projects related to agricultural practices in southern Arkansas. The predominantly black institution sought to find its research niche by assisting farmers with limited capital in upgrading operations. In 1977, Congress enacted the Food and Agriculture Act that incorporated the provision to enable the 1890 institutions to begin long-term projects in agricultural research. In Arkansas, the new, rapidly growing area of aquaculture, including catfish, bait fish and crayfish, was identified as a major research initiative. A complex of 72 experimental ponds, a fish hatchery and a five-acre water reservoir to research improved breeding and disease control for fish was built at the University of Arkansas at Pine Bluff.

In February 1973, E.M. Cralley retired as director of the Arkansas Experiment Station. Cralley was replaced by Lloyd O. Warren, an Arkansas native, a graduate of the University of Arkansas and Kansas State University and an entomologist who had extensive experience in the development of the cotton insect scouting program and the Arkansas Insect Survey and in forest entomology. Warren had worked with Cralley in the construction, staffing and management of the Virology Laboratory.

Higher energy costs combined with a general rise in the cost of all oil-based products placed farmers in an economic vise. Warren believed that dynamic research was essential to assist the farmer in dealing with this new challenge. Greater mechanization, improved varieties and more effective pest control were necessary for future progress. Land hitherto considered worthless or worn out for agricultural purposes would have to be reevaluated and brought into production. Problems of mineral build-up and contamination by fertilizers, pesticides and suspended matter due to rice irrigation meant reexamination of soil chemistry in the Grand Prairie aquifer. The Experiment Station tested the post-emergent herbicide Dinoseb to deal with cockleburrs, a serious weed threatening cotton and soybeans. Experiment Station scientists discovered that zinc seed coating could prevent zinc deficiency. The release of 'Lee 74' soybean, 'Quapaw' and 'Stripper 31' cotton and 'Bonnet 73' rice increased yields of those crops.

Joe McFerran, who had been on the faculty since 1946, played a major role in the continuing revival of truck farming in Arkansas, developing disease-resistant varieties of tomato, okra and turnip greens. In 1976 he was named chairman of the Southern Region of

Entomologist Lloyd O. Warren was appointed Station director in 1973. During his 10 years as director, he was a strong proponent of interdisciplinary research. He also coordinated the expansion of the three branch research units at Stuttgart, Monticello and Keiser into research and extension centers and managed the transition of agricultural research programs at UA-Pine Bluff and of forestry research at UA-Monticello into the Station. He became emeritus professor in 1986.

the American Society for Horticultural Science. Another plant breeder, George A. Bradley, who investigated mineral nutrition of vegetable crops, was named Fellow of the American Society for Horticultural Science in 1975.

The new field of ornamental horticulture was initiated by A.E. Einert who started programs to assist nursery and horticultural industries in Arkansas. In 1974, Gerald Klingaman was hired as an expert on woody ornamental plants to assist Arkansas commercial nurseries.

The worsening of the energy crisis resulted in the National Science Foundation forwarding $1 million in "pass though" grants to the Agricultural Research Service for investigations in solar energy applications in agriculture. In 1975, the Cooperative Research Service, working with the Energy Research and Development Administration, funded the use of solar energy in crop drying, green houses, rural residences and food processing. Agricultural engineers developed an automatic control system to allow farmers to store seed as the cotton was picked. In this fashion, pickers could operate at full capacity without the need for new, expensive trailers.

In 1975, the Experiment Station leased 46 2/3 acres of land south of Kibler to replace Whiteside Farms near Van Buren as the Vege-

Horticulturist Joe McFerran (left) and plant pathologist Jack Goode (right) worked together to develop new disease-resistant tomato varieties, including the famous 'Traveler' varieties.

table Substation. In 1982, the site was purchased, along with another 20 acres. The staff studied cultural and nutritional factors impacting quality and yield of vegetables such as sweet potatoes, cantaloupe, green beans, watermelon and southern peas. New cultivars in tomatoes, sweet corn, okra, southern peas and cucumbers were developed as well as two spinach cultivars, 'Ozarka' and 'Greenvalley', which were resistant to white rust and fusarium diseases. Experiments in small grains, wheat, soybean and alfalfa were also initiated.

In 1975, the Delta Branch Station at Clarkedale was formally created. The Delta Branch had been a substation of the Cotton Branch Experiment Station at Marianna since 1939. The 222 acres of owned and leased land was devoted to such crops as soybeans, small grains, cotton, winter annual legumes, corn, grain sorghum, rice and peanuts in three different soil types, Sharkey Clay, Dubbs and Dundee sandy silt loam. Today projects include pest and weed control, soil water management and use of cover crops for soil build-

The Vegetable Substation was established near Van Buren in 1948 and was later relocated to new facilities near Kibler. Breeding programs at this location have involved corn, tomatoes, okra, spinach, mustard greens, cucumbers, cantaloupe, watermelon, southern pea, soybeans and small grains. Pictured is one of the first mechanical greens cutters used at the Station. (Photo courtesy Dennis R. Motes.)

ing. Presently, special efforts are being made in examination of the root-soil microenvironment and its relation to plant growth.

Congress in 1975 began to target grants to spur on specific programs such as beef and pork production, food and nutrition. In 1976, Congress launched the National Pesticide Impact Assessment Program to broaden ongoing programs to promote integrated pest management with chemical and biological controls, post-emergent and cultural techniques in order to limit the adverse economic impact of pests and pathogens on the range of farm crops. The Arkansas Station served as the lead station in the Southern Region during the early years of this effort.

The Agricultural Experiment Station renewed its efforts to examine the root causes for the failure of the Ozarks to revive economically. On one hand, the indigenous populations possessed limited capital, infertile soil, a poor transportation network and a pessimistic outlook on the possibilities of economic advancement. On the other hand, new factories in the Ozarks resulted in new patterns of employment and income, and the cheap land, salubrious climate, artificial lakes, low taxes and low crime rate had made the Ozarks attractive for recreation and new retirees. "Back-to-the-landers," retirees and returning residents caused a major jump in the population of communities in northern and northwestern Arkansas. New immigrants had higher expectations of social services, especially medical care, than those already existent. On the other hand, the educational demands of the indigenous population were opposed to retirees who no longer had children in the school system. Sociologists tried to discern the problems resultant from the interplay between these two highly divergent cultures.

Sociologists also explored the impact of retirees on Arkansas' limited health resources. The aging rural population combined with the establishment of retirement villages and homes gave Arkansas one of the highest percentages of elderly in the nation.

As part of the expanded interest in the status of the elderly, the Experiment Station was awarded a series of grants in the late 1970s to examine the impact of nutrition programs on the aged (Bulletin 839). Researchers Michelle Davis Fryar, Donald Voth and Mary Jo Schneider concluded that the federally sponsored nutrition programs did not measurably improve the health status of the elderly or reduce their hospital visits, but the programs did bolster their self-image and social contacts and assist them in dealing with problems. Follow-up studies indicated the distaste of the elderly for nursing homes, but the economics of home health care were influenced by the rural distribution of the population, which raised transportation costs. Diana M. Danforth and Donald Voth in 1981 confirmed that Arkansas could anticipate a continued in-migration of retirees to the Ozarks due to recreational benefits and the return of out-migrants from the 1930s and 1940s (Bulletin 855). This aging population would make increasing demands on the state's inadequate social services.

In eastern Arkansas, rural sociologists confronted different problems. Poor education, low skill levels and isolation made the outlook for displaced agricultural workers unpromising. Rural sociologists cautioned that industrialization was not practical without an upgrading of the vocational and educational facilities. Mary Jo Schneider and Bernal Green recommended an adequate youth jobs program to counter low motivation and urged education as the key to stem out-migration.

The Experiment Station continued to work with governmental agencies as an independent investigator of changes in rural America. Home economist Cora McKown surveyed the housing needs of residents in Newton County. Gordon Morgan surveyed the economic needs of Lakeview, a small black community located in eastern Arkansas. Robert Shulstad and T.G. Sawyer explored the need for additional public recreational usage on Beaver Lake.

Rural sociologists delved into the problems of the perceived need for increases in funding levels for agricultural research. With a decline in the number of active farmers as well as persistent surpluses, the Experiment Station had to reassure the general public of the vital nature of ongoing work. Because of the paucity of public service funding, many Arkansans believed that roads, recreational facilities and schools were of more pressing importance than additional state incremental funding of the Experiment Station. Even given that agriculture was the leading state economic sector and its tax receipts furnished a major share of the tax revenue necessary to fund other areas of the economy, the primacy of agricultural pursuits was challenged.

Federal agencies involved in Medicare and Medicaid considerably upgraded the health of many Arkansans. Despite these advances, some 40% of Arkansas remained below the poverty level. Agriculture still provided a major source of income for the rural population of Arkansas, often supplemented by factory labor in a nearby town.

With rapid appreciation in land values, farmers found that passing down farmsteads from generation to generation was difficult due to estate taxes. Assisting farmers in estate planning became an important aspect of the perpetuation of the family farm. To assist families with this important economic activity, J. Martin Redfern, Henry Meenen and John R. Graham wrote for farmers detailed instructions for dealing with federal and state inheritance taxes in order to minimize the tax bite (Bulletin 771). Farmers were instructed in the necessary measures to lower tax liabilities.

Advances in agricultural techniques had become an increasingly important factor in foreign affairs. The American public viewed a world that could feed itself as one that would be less susceptible to

VARIETIES AND STRAINS DEVELOPED BY THE ARKANSAS AGRICULTURAL EXPERIMENT STATION IN THE 1970s

AGRONOMIC CROPS

Alfalfa	
Victoria	1971
Corona	1979
Cotton	
New Rex	1976
Grain Sorghum	
AKS 618	1973
Oats	
Bob	1977
Rice	
Nortai	1972
Bonnet 73	1973
Nova 76	1976
Mars	1977
Soybean	
Mack	1971
Lee 74	1974
Hood 75	1975
Wheat	
Doublecrop	1975
Rosen	1979
White Lupine	
Hope	1971

HORTICULTURAL CROPS

Blackberry	
Cherokee	1974
Comanche	1974
Cheyenne	1976
Grape	
Venus	1977
Okra	
Lee	1978
Southern Pea	
Bush Purple Hull	1973
Elite	1978
Spinach	
Bounty	1970
Strawberry	
Cardinal	1974
Comet	1975
Tomato	
Traveler	1970
Traveler 76	1976
J. Pope	1977
Turnip	
Improved Crawford	1974

James E. Martin was appointed vice president of the Division of Agriculture in 1975. In 1980 he was appointed president of the University of Arkansas and served until 1984.

communist influence and political unrest. Starting in 1976, the United States Agency for International Development gave special grants to encourage cooperation and exchange of technical information. The Arkansas Experiment Station had contacts with projects in Panama, Haiti and Bolivia. Students from all over the globe were able to study in Arkansas. Private foundations such as Winrock International, financed by the Winthrop Rockefeller Foundation, became sources of funding for areas of specific interest such as raising goats.

By 1975, the pace of agricultural research in Arkansas was accelerating due to the utilization of genetic and biochemical innovations. The soybean breeding program under C.E. Caviness produced 'Hood 75', which was resistant to rot and matured 8 to 10 days earlier than 'Lee 74'. The shortening of the growing season permitted double cropping with soft early red winter wheat or grain sorghum and allowed farmers a possible major increase in income. In 1975, F.C. Collins and J.P. Jones developed 'Doublecrop', a new variety of soft red winter wheat that replaced 'Arthur' wheat for planting after cotton, soybeans and grain sorghum.

Berry breeding and production were major interests of the Experiment Station as farmers needed a high-value crop that could be grown in the upland region. Agricultural engineers worked with

In the early 1960s, horticulturist James N. Moore started a plant breeding programing in small fruits that has led to the release of six varieties of blackberries, three strawberries, including 'Cardinal', and four varieties of seedless table grapes.

Blueberry Equipment Inc. of South Haven, Michigan, on the strawberry picker and built a continuous flow machine to clean and grade fruit. Strawberry varieties 'Cardinal' (1974) and 'Comet' (1975) were released; the blackberry breeding program released 'Cherokee' and 'Comanche' (1974) and 'Cheyenne' (1976); and a new tablegrape variety, 'Venus' (1977), was released.

In 1976 with the election of Jimmy Carter, the so-called "New Agenda" in agriculture became an issue. Bob Bergland, the new Secretary of Agriculture, was an advocate of emphasis on social implications of farming. The New Agenda evolved out of complaints about industrialized agriculture including chemicals, soil erosion, the growth of farm size, difficulties of securing adequate loans and the decline of the quality of life in rural America. The new administration favored safeguards on the application of biotechnology. Congress in 1977 enacted the Food and Agriculture Act, which sought to place all the multiplicity of demands under one umbrella.

The Federal Home Administration and the Small Business Administration declared a moratorium on foreclosures as well as a $4 billion emergency loan program. The Carter administration made the argument that economic aid to farmers created a healthier export market leading to higher farm programs. Unfortunately, the grain embargo in 1980 in the wake of the Russian invasion of Afghanistan allowed for other agricultural nations to cut into Arkansas' market for soybeans.

By the end of the decade, the Carter administration created the position of Assistant Secretary for Conservation, Research and Education with an overview of the Cooperative State Research Service. Outside experts were invited to work with Station scientists and administrators on ways to improve research. This procedure was modified to include an "in-station" peer review process for every research proposal. Moreover, non-profit research organizations were allowed to bid on new research initiatives.

The Carter administration favored regionalization of the Agricultural Research Service working with the state-based scientists. To promote greater coordination among the various research entities, the Agricultural Research Policy Advisory Board was started to chart and direct future research priorities apart from "production agriculture" and to include over production, soil erosion, rapidly escalating farm prices and costs as well as the impact of export policy. Indicative of the various interest group demands, the bill encompassed 15 subject areas including the use of natural fertilizers and fuel efficiency.

A Joint Council on Food and Agricultural Science was formed to advise on research priorities. Moreover, the National Agricultural Research and Extension Users Advisory Board was created to articulate the demands of the various constituencies affected by farming. Congress created a new category of funding referred to as a competitive grant to develop "scientific breakthroughs" in solar power and livestock and poultry research. In response to civil rights activists, more funding was also to be diverted to the black land-grant schools, or 1890 colleges, such as the University of Arkansas at Pine Bluff.

Secretary Bergland also directed research money to assist the family farmer, defined as having less than $20,000 in annual gross sales. Activities to assist the beleaguered small farmer were compiled in a series of national meetings held in 1978 by the U.S. Department of Agriculture to ascertain farmers' needs. In addition, an Agricultural Mechanization Task Force was charged with the responsibility of evaluating the economic, social, environmental and labor displacement of farm machinery. As Bergland viewed it, the thrust of research would be the entire spectrum of farming-related activities, including environmental impact.

The New Agenda in agriculture was reflected in the creation of the Small Farms Research Center at Booneville. The genesis of the project was a study conducted in 1976 to propose a new use of the land of the Arkansas Children's Colony and Arkansas Tuberculosis

Sanitarium. After an initial study by the U.S. Department of Agriculture and University administrators, suggestions were made for utilization of the acreage. In 1979, Congress authorized a research program at the Booneville facility for animal forage systems and production of fruits, nuts and vegetables to serve the southwestern region. In 1980, the Small Farms Research Center started renovation and improvement of 1600 acres for work in vegetables, fruit and nut crops, soil and water conservation, rural characterization, aquaculture, animal entomology and systems research with the goal of the research being the preservation of the family farm.

In 1978, agricultural research at Arkansas State University, Jonesboro, became a part of the Arkansas Agricultural Experiment Station system in order to avoid duplication of ongoing programs. Dr. George Berger, dean of the College of Agriculture, concentrated on development of soybean cultivars adapted to soils with a high salt content, extreme pH levels or low fertility as well as development of new cultivars that could be used in double cropping and narrow row spacing and identification of improved levels of resistance to the soybean cyst nematodes. Arkansas State University has also conducted experiments on the efficacies of various fungicides for wheat as well as the adaptability of rapeseed to Arkansas conditions.

Regional cooperation intensified to maximize the interchange of scientific data. In 1977, Arkansas participated with 12 other Southern states trying to breed new germplasm to help breeders develop disease-resistant varieties of both horticultural and agronomic crops. The private sector funding of individual research projects had become increasingly important for the Station. Experiment Station personnel worked closely with agriculture-related businesses to appraise products for efficacy and health as had been the tradition of the Station from its inception. The Southwest Research and Extension Center at Hope and the Livestock and Forestry Station at Batesville worked with Nekoosa Paper, Georgia-Pacific Corp., Potlatch Corp., International Paper Co., Weyerhaeuser Co. and Ross Foundation as well as the Arkansas Science and Technology Authority. In 1978 C.D. Cotterman tested commercial biochemical stimulants to see if they altered the metabolism of cotton plants in favor of increased seed cotton yields. The Experiment Station published performance examinations of commercial seeds.

The scope of Agricultural research continued to rapidly expand in the 1970s in response to increasingly complex agricultural problems, Congressional mandates and continued economic pressure. B.R.

Poultry scientist Roy Gyles participated in research beginning in 1959 that led the broiler industry to change from pea comb to single comb male parent lines. He was the first to measure the feed converting ability of individual candidate breeder turkey toms in single floor pens and to demonstrate the presence of variation for feed converting ability among turkey toms of similar body weights. He later assisted in the adoption of a similar program for the selection of candidate cockerels in commercial male parent meat-type chickens. Gyles conducted pioneering research that demonstrated that a genetic mechanism is involved in the spontaneous regression of a malignant tumor in animals and that the genetic mechanism acts through the immune system.

Wells and Wayne Sabbe began tests on the need for sulfur in rice growing. The Experiment Station constructed a solar-heated chicken house to determine if major economies could be achieved in the face of soaring propane prices. C. Roland Mote and Richard Estes tried to devise a method of disposing of chicken carcasses by use of heated septic tanks. 'Corona', a new alfalfa variety, was released to provide superior forage and resist phytophthora root rot. In 1979, the Experiment Station released 'Rosen', a high-yielding, soft-red winter wheat. This new semi-dwarf wheat variety, named in honor of H.R. Rosen, was resistant to major wheat diseases in Arkansas. Ronald Day and M.V. Meisch tried dispersal of mosquito fish in the rice field as a supplementary form of insect control. Robert Frans, E.J. Matthews and L.R. Oliver conducted an extensive study of various chemical and natural methods of weed control. The Experiment Station played an active role in national and regional efforts to ascertain safe levels of herbicides.

The poultry industry in Arkansas was a major beneficiary of research results as the state gradually became the national leader in chickens. In 1940 it took about 80 days and 4.12 lb of feed to produce a pound of broiler; by 1950 it took 3.3 lb; and in 1963 it took 2.5

lb of feed. By 1977, a 4-lb bird could be brought to market in 55 days on 1.9 lb of feed for each pound of chicken, and by 1989 the time had been further cut to 42 days. Faster gains, mechanized facilities and better housing reduced labor costs. The average number of eggs laid per hen in 1950 was 118 compared to 235 in 1975. The poultry industry became the single largest employer in the state.

Forest research at the Southwest Research and Extension Center, started in the 1920s, had been dealing with species performance, plantation management and cultural practices. In 1947, F.M. Meade started intensive management of wood lot and pine and hardwood research. From 1970 to 1974, H.A. Holt investigated techniques of vegetation control as a tool in site preparation as well as screening of various herbicides. Lloyd O. Warren explored the impact of the pine top moth. A.T. McDaniel started selective breeding of Christmas trees under Arkansas conditions.

By 1980 the scope of agricultural research had gone far beyond the parameters first envisaged by the initiation of the Hatch Act. However, the next decade with its innovations in biogenetic engineering promised to totally reshape the face of agriculture.

Chapter VIII

The Experiment Station Today

The decade of the 1980s has offered exciting new directions in agriculture. The conflict among the emphasis of the "New Agenda" on environmental protection, the needs of the small farm and the traditional emphasis on higher productivity was resolved with a renewed emphasis on higher yields with the Reagan administration. In 1981, the National Agricultural Research Extension and Teaching Policy Act reinforced the continuing commitment to state agricultural research station programs as well as the ongoing regional and national programs. In order to restore America's role as a food exporter, increased production once more became the top priority. However, Congress retained its special programs and cooperative grants with funding for basic research into biological studies on nitrogen fixation, photosynthesis and animal and human nutrition as well as water and soil chemistry. The Joint Council and Users Advisory Board was expanded to continue to provide input into the nation's planning processes. Renewed emphasis was to be placed on improvement in range management and agriculture. The Agricultural Experiment Stations were directed to work with the Agency for International Development to assure agricultural advances in Africa, Asia and Latin America. The Experiment Stations were placed under the Assistant Secretary of Science and Education who was charged with the "rigorous evaluation" of proposals to avoid unnecessary duplication of efforts.

The Reagan administration also wished to break with the Carter agricultural policy. During 1980, a considerable amount of grain had been moved into storage due to the surplus created by Carter's Russian grain embargo. As a result, the administration decided on a free market approach to acreage control, depending on the free market to absorb the surplus. By 1981 crop acreage was the highest in history with a 600% increase in wheat acreage and a 300% increase in rice acreage over 1971 levels. Arkansas was the leading rice producer in the nation with 69 million cwt, 38% percent of the nation's output.

The increase was not altogether welcome. America found itself with a "mountain" of surplus grain by 1984. In a reversal of policy, the government was forced to return to major set-aside programs with some 77 millions acres not farmed. Unfortunately, unanticipated drought resulted in a rapid surge of grain prices, adversely impacting beef producers. Overall, farmers trapped by increased costs, more competitive international markets, an inflated dollar and high interest costs labored in an adverse economic environment.

The international market had become an important determinant of the future of Arkansas agriculture. Agricultural economist Eric Wailes created models to determine world future demand for rice. With the elimination of the allotment program to restrict rice acreage, the overseas market became essential to Arkansas producers as the nation's leading grower of rice. World-wide fluctuations in demand impacted farming in Arkansas; consequently, the farmers' need for accurate market analysis became paramount.

In 1980, the University dedicated the Altheimer Laboratory on the University Farm to study plant growth under laboratory conditions. A second facility located at the University of Arkansas at Pine Bluff was built to target programs such as aquaculture, sweet potatoes, rabbits and poultry diseases. The Northeast Research and Extension Center at Keiser was opened to work on alfalfa, cotton and soybeans. The constraints on row crops in the Delta put an emphasis on the most cost-effective usage of acreage.

John W. Goodwin was appointed vice president of the Division of Agriculture in 1981. He resigned in 1986 to return to teaching and research in the College of Agriculture and Home Economics.

In 1982 as farmers were involved in growing "fence post to fence post," Billy Herrington and Robert N. Shulstad developed an economic feasibility study on conversion of Delta woodland and pasture to cropland (Bulletin 858). In their examination of the practicability of bringing new land into production, they cautioned about soil erosion, destruction of water life and water run off and the need for wind breaks. The subsequent drought in 1988 demonstrated the prescience of their concerns.

The Arkansas Experiment Station continued its traditional lines of inquiry into fertilizer application, herbicide efficacy and safety, cooperative beef-bull performance, integrated pest management, apple and peach trials and calf weaning. The cooperative testing program provided cattle breeders with necessary information to select animals with superior breeding value. From 1962 to 1981, the average gain increased by about 50% in gross feed efficiency and contributed to the expansion of the cattle industry as well as the upgrading of cattle hybrids.

Major research objectives included corn hybrids that could resist maize dwarf mosaic. C.E. Caviness released a new soybean variety, 'Jeff', which was resistant to the cyst nematode. J.L. Bowers, M.J. Goode and W.A. Sistrunk released a new cream type southern pea cultivar dubbed 'Erectset' that had a high level of bacterial blight resistance.

James N. Moore developed the third in the series of strawberry varieties, 'ArKing' (1981). In addition, Moore also developed new hybrids of seedless table grapes, 'Reliance' (1983), 'Mars' (1984) and 'Saturn' (1988), that have become a major new sector for Arkansas growers. Other new varieties released included blackberries--'Shawnee' (1984) and 'Choctaw' and 'Navaho' (1988)--as well as peaches--'Allgold' and 'Goldilocks' (1984)

Joe McFerran, M.J. Goode and W.A. Sistrunk collaborated on a new okra variety, 'Jefferson'. T.E. Morelock, M.R. Hally and W.A. Sistrunk released a new mustard variety, 'Slobolt'. 'Nelson', an early-maturing, high-yielding, soft red winter wheat named after Martin Nelson, was developed to assist farmers who double cropped. The Station released two tomatoes, 'VF Pink', disease resistant (1982), and 'Arkansas 7985', resistant to fruit rot (1981).

The Experiment Station continued in its investigation of the general concerns of rural Arkansans. T.R. Rokeby and C.C. Griffin studied the new techniques of heat storage for the solar-heated chicken house. Food and nutrition scientist Esther Glover studied the leisure and physical activity of low-income, elderly Arkansans. Mike Miller and Don E. Voth established a statistical basis for estimates of incidents of hypertension among rural Americans. Bernal Green continued his in-depth analysis of the matrix of poverty in the Ozarks.

In 1980 Arkansas experienced a drought that reduced the important soybean yields. As a consequence, farmers and researchers showed a renewed interest in the efficacy of irrigation. H.D. Scott and J.A. Ferguson studied the correlation of yield to soil temperature and moisture. Overall acreage under irrigation in the state has been increasing at a rate of 34,000 acres per year over the last 40 years. Today about one-third of the farmers in eastern Arkansas

VARIETIES AND STRAINS DEVELOPED BY THE ARKANSAS AGRICULTURAL EXPERIMENT STATION IN THE 1980s

AGRONOMIC CROPS		**Grape**	
		Reliance	1983
Cotton		Mars	1984
Arkot 518	1987	Saturn	1988
Lespedeza		**Mustard**	
Marion	1989	Slobolt	1980
Rice		**Okra**	
Bond	1983	Jefferson	1981
Newbonnet	1983	Jade	1988
Tebonnet	1984		
Katy	1989	**Peaches**	
		Allgold	1984
Soybean		Goldilocks	1984
Jeff	1981		
Narow	1983	**Southern Pea**	
Lloyd	1987	Erectset	1981
		Epoch	1984
Wheat		Encore	1986
Nelson	1982		
Keiser	1987	**Spinach**	
		Ozarka	1980
HORTICULTURAL CROPS		Green valley	1980
		Fallgreen	1987
Blackberry			
Shawnee	1984	**Strawberry**	
Choctaw	1988	ArKing	1981
Navaho	1988		
		Tomato	
		ARK 7985	1981
		VF Pink	1982
		Ozark Pink	1988

irrigate their fields, and Arkansas is eighth in the nation in amount of land irrigated.

The Experiment Station often found itself being regarded as the dispassionate investigator of politically charged questions. A particularly sensitive concern was human exposure to 2,4-D during its aerial application on broadleaf weed species. The initial findings of Station scientists confirmed the assertion of chemical companies that the product was safe if properly utilized. Another public policy question addressed by the Experiment Station was the disposal of the more than 33 million broilers that died yearly in Arkansas prior to maturity. Richard Estes and C. Roland Motes examined the possibility of burying carcasses in heated septic tanks. In dealing with the complex method of establishing parity price for milk support, Calvin Berry and Eric Wailes made computer projection profiles to ascertain the optimal price point.

The work of the Experiment Station was fraught with examples of unanticipated consequences of early agricultural advances. For instance, Arkansas stock raisers utilized fescue grass as a drought-resistant, cool-season forage crop. Unfortunately, tall fescue became

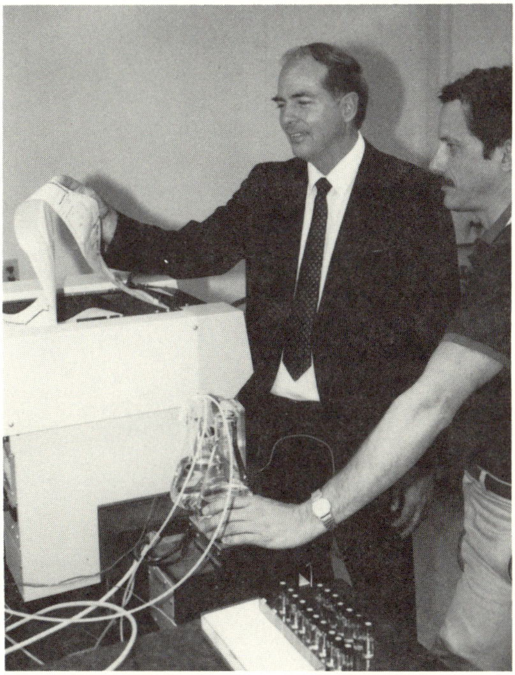

During the 1980s, agronomist Terry Lavy, pictured here on left with research associate John Mattice, conducted studies measuring pesticide exposure of humans applying or working near pesticide areas. Results of his work are used in field worker training and in risk assessment by government regulatory agencies and industry.

Agronomist Don Brown modified a research tool called a duodenoscope to photograph and then video tape the root system of plants growing in the ground.

toxic to cattle and resulted in beef animals with reduced gains, decreased feed and impaired reproduction. T.S. Nelson, L.B. Daniels, E.L. Piper and J.N. Beasley started cooperative research on isolating the toxic substance in fescue.

In 1983, the outlook for farmers darkened. The confluence of high interest rates, stagnating land values, an over-valued dollar, the flattening of international demand and the high cost of agricultural input all contributed to a low profit margin for farmers. Even a one-cent differential in broiler prices resulted in a multi-million dollar difference in income. The Arkansas Agriculture Experiment Station studied the use of paper mill primary sludge for broiler litter as a cost control device. New sources of protein, such as sunflower-meal extraction, were explored by O.T. Stallcup. Researchers utilized underground cameras to study root maturity of soybeans and the influence of soil properties. To lower pickling costs, R.W. Buescher utilized a hollow fiber ultrafiltration system for the recycling of brine utilized in pickling.

Governmental agricultural policy, with its emphasis on farm subsidies, had a major impact on farm incomes. Don Voth, W.A. Halbrook, D.D. Chapman and R. Renfro analyzed the operation of small farms in western Arkansas. C.R. Berry monitored the impact of price support policy on dairymen. Daniel Capstick and Mark Cochran worked on programs to assist farmers in developing procedures for making planting choices.

Pest management in an era of environmental concern became increasingly difficult. In 1982, the National Integrated Pest Management Coordinating Committee was established to exchange information regarding ongoing experiments. In soybeans, two fumigants that controlled soybean cyst nematode were removed from the market. R.D. Riggs and C.E. Caviness concluded that in the future, nematodes would be controlled by a rotation program utilizing nonhost crops, resistant and non-susceptible cultivars. Congress asked the Experiment Stations to explore the possibilities of "organic" or non-chemical farms in order to avoid the perceived dangers of herbicides and insecticides. Congress also enacted the Small Business Innovation Development Act to direct ongoing technological development of small firms involved in marketing innovations. Added product value was seen as essential to future income growth for small farmers. Farmers markets, direct marketing and pick-your-own berry farms bridged the gap between the producer and the consumer.

Energy efficiency and reducing the use of petro-chemicals were considered essential in the face of low profit margins. The Experiment Station continued studies on improved insulation of poultry houses. L.H. Hileman and T.E. Morelock explored the use of cotton gin trash compost as fertilizer for lawns, gardens and greenhouses. The Experiment Station continued to perfect the mechanical blackberry harvester in the face of rising labor costs. J.R. Morris discovered a method for hastening the ripening of fruit to allow for once-through harvesting. In lieu of readily available hand labor, weed control with preplant, pre-emergent and post-emergent herbicides was necessary for successful harvests. The continuing improvements in rice varieties led to the release of varieties 'Bond' (1983), 'Newbonnet' (1983) and 'Tebonnet' (1984).

In 1983 Lloyd Warren stepped down as director of the Experiment Station to be replaced by Preston La Ferney. In his farewell message, Warren warned that agricultural research was under siege. With a reduced constituency and no increases in funding, Warren feared a division of the research thrust between the U.S. Department of Agriculture's basic research and the state Experiment Station's applied research. He predicted that this approach would undermine the traditional cooperative research that underlay the state-based Experiment Station philosophy.

Agricultural economist Preston E. La Ferney, who served as associate director of the Station, succeeded Lloyd Warren as Station director in 1983. In 1987 he became vice president for the University of Arkansas Division of Agriculture.

The entire field of agricultural research seemed poised for a new revolutionary direction with the advent of biogenetic research. Starting in 1981, the Experiment Station requested money to commence experiments in gene and tissue manipulation to create new hybrids. In 1984, J.W. King and F.H. Huang began experiments with bermudagrass to breed mutants with superior characteristics and enhanced resistance to temperature variations. Huang also worked with the Arkansas Forestry Commission to develop the Scotch pine tree from root cuttings.

Rice cultivation had long remained a major research interest, especially as rice producers had been in a profit squeeze. A combination of poor growing conditions, higher energy costs, shifts in the Asian markets from importers to exporters, the falling water table on the Grand Prairie and new rice diseases all contributed to problems of rice growers. In 1983 the Rice Research Board asked the Experiment Stations to conduct additional verifications trials. An advisory committee of ten scientists analyzed data to assess needs. The objective was to test varieties and inputs to furnish farmers with all vital market information. A program of fertilizers, planting and cultivation information was furnished to farmers along with foundation seed.

The Home Economics department worked on tests of garment comfort properties. Diana Cone designed special clothes with velcro closures to assist women who suffered from arthritis.

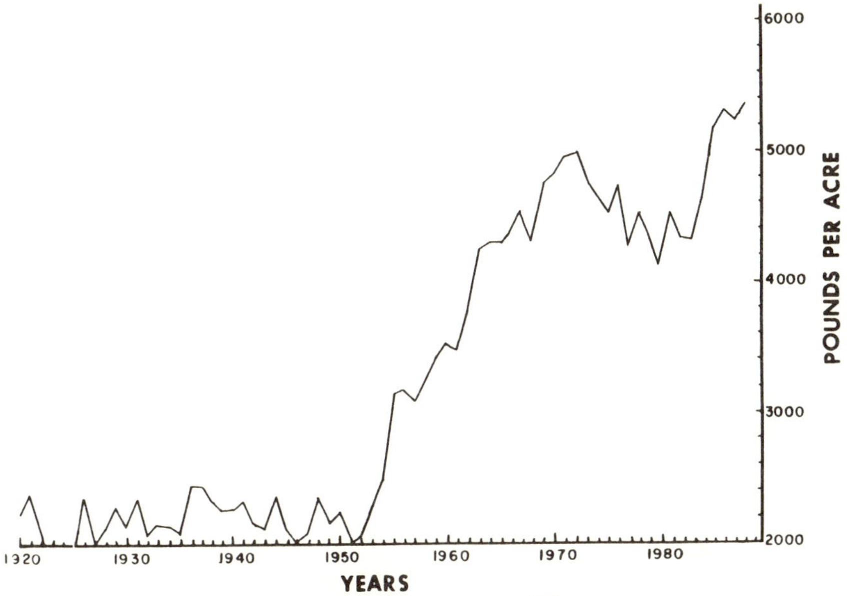

Average rice yields in Arkansas, 1920-1988.

In food processing W.A. Sistrunk and A.R. Gonzalez in 1983 undertook a study of the quality of mechanically harvested snapbeans, concluding that snapbeans had to be harvested and processed in the same day to avoid the chemical and physical changes that took place during the period of holding at a high temperature and resulted in a decrease in quality. In addition, proper blanching was important to maintain quality.

The Agricultural Experiment Station as a major state research entity was utilized in environmental studies. Arkansas scientists monitored acid rain as part of the National Atmospheric Deposition Program. In their readings, Scott Beasley and Alfredo Granillo observed variations in the pH values of rain that could have an impact on plant-growing capacities. The Experiment Station also was charged with disposal of nuclear waste from an experimental closed fast breeder reactor.

The diversity of crops in Arkansas reflected itself in the continuing growth of the beef industry. The Arkansas Beef Council helped finance an investigation into the reasons for the toxicity of fescue grass, the major cool-season forage in Arkansas. After exhaustive research, L.B. Daniels, E.L. Piper and B.J. Hankins found a correlation between seed-borne fungus and the adverse reaction by cattle. The department of animal sciences offered to test pastures for fescue toxicity as well as make available fungus-free varieties of fescue.

In further beef cattle studies, A.H. Brown and C.J. Brown worked with researchers at the Southwest Research and Extension Center at Hope to continue cross-breeding experiments comparing maternal abilities of beef cows from 1979 through 1982. Four herds were established and bred with Brangus and Simmental bulls. By 1985, approximately 50,000 farmers were producing 1 million feeder calves, mainly crossbreeds. Tests of dairy products included study of factors influencing keeping quality and flavor of milk, butter and cheese and factors limiting shelf life.

A new agricultural endeavor in Arkansas--raising rabbits--became the new research area in the 1980s. L.B. Daniels worked with rabbit growers to see if coastal bermudagrass treated with anhydrous ammonia could be utilized as a substitute for more expensive alfalfa. T.S. Nelson and Daniels worked on a basal diet for rabbits to determine optimal growth rations. The industry offered an opportunity for rural families to supplement their incomes.

Research in veterinary science concerned Johne's disease, bovine infertility, the regulation of immunity in domestic animals and birds, neonatal diarrhea in cattle, *Mycoplasma synovia* infection in poultry and gastrointestinal parasites of ruminants. Studies of diseases included Marek's disease, infectious bursal disease and Newcastle disease.

Arkansas agriculture, like that of other major farming states, was hurt by the residual effects of the embargo, new international agricultural exports, the export drive of the Common Market and the reduced buying power of Third World nations due to their already heavy debt. High-interest loans contracted during the inflation of the late 1970s combined with declining collateral values of farmland forced many American farmers to face the possibility of foreclosure.

Congress sought to deal with this economic problem by the Food Security Act of 1985. This bill tried to encompass both traditional concerns of even higher productivity and the "New Agenda" emphasis on environmental awareness. Congress delineated research areas including "biological methods of weed, disease and insect control . . . legume crop rotation, the use of green manure, animal manure and municipal waste" and mandated a substantial increase in funding for the biotechnology of recombinate DNA. The vision of superior genetic breeding seemed to offer an attractive alternative to chemical inputs that were perceived as environmentally damaging. However, fears of an accidental creation of a "Frankenstein" germ or plant led to the inclusion of a standardization of risk assessment procedures and a temporary ban on the field testing of laboratory findings. In 1986, the Biological Sciences Coordinating Committee was created to institute regulation of biotechnology.

Congress also directed the Experiment Station to continue to provide research information to small farms. Moreover, Congress requested that the Experiment Stations research "quality enhancing post-harvest technology." The concept of value-added product development was already being successfully pursued by Arkansas' integrated poultry producers with frozen, pre-cooked dinners. Finally, Congress created the International Trade Development Center to assist in opening and servicing foreign markets.

In revising the legislation guiding the Experiment Station, Congress perpetuated Title XIV of the National Agricultural Research Extension Users Advisory Board until 1990. The Joint Council worked with state Station directors and the Assistant Secretary of Agriculture to serve as a conduit of opinion on research direction. A sensitive reading of public opinion was necessary in the face of the desire of the Office of Management and Budget to cut domestic federal funding. The stringencies of federal funding were reflected in the budget mix of the Arkansas Experiment Station in which 63% of the money was derived from state funding, 19% from federal sources, 8% from sales of foundation seed and commodities and 10% from grants. With the agricultural states suffering from revenue shortfalls, it was becoming difficult to maintain the entire range of research effort of the Station. To quote C. Dayton Steelman, associate director from 1983 until 1987, the Experiment Stations were at "crossroads of significant proportions." Sponsored research would in the future play a major role in determination of new directions. The continuation of the check-off system in soybean, rice and cotton --the levying of fees based upon per-bushel harvest to help defray the cost of ongoing research--became a matter of controversy.

In order to define future areas of interest, the state Experiment Stations in 1985 created their own Research Planning Subcommittee to assure the niche of the state-based entities. The Stations issued a report entitled *Research Perspective Papers* as the consequence of a four-day symposium in Washington.

The scientific evolution of Arkansas farming continued apace. Despite a major reduction in cotton acreage from 2 million acres in

In 1987 Gerald J. Musick became the third entomologist to serve as Station director. Prior to his appointment, he was head of the department of entomology.

1953 to 290,000 acres in 1983, yield had increased from 292 lb/acre to 700 lb/acre by the mid 1980s. Arkansas' yield was approximately 5% to 13% higher than the national average.

In 1987, Preston La Ferney was promoted to Vice President for Agriculture and was succeeded as director by Gerald Musick. Musick assumed the mantle of director at a time when low domestic prices and an overvalued dollar had a major adverse impact on agricultural exports. The Station sought to help producers restore profitability by implementing production systems and by seeking ways to minimize costs while "increasing productivity and product quality." The Experiment Station formed the Arkansas Cotton Production Marketing Research Group to solve the problems that required immediate attention. The cotton research group sought to integrate physical, biological, economic and management knowledge into the production system.

Today with advances in genetic research, scientists have the potential to develop varieties with disease and pest resistance hitherto impossible by conventional means. M.J. Goode in plant pathology and T.E. Morelock in horticulture developed a new spinach cultivar, 'Fallgreen', to restore production of that vegetable in the Arkansas River Valley once threatened with destruction due to disease. Joe M. McFerran, S.J. Scott and Rose Gergerich have contributed to major improvements in the tomato industry in Bradley County. Research by F.H. Huang has revealed a new technique of tissue culture that accelerates the process of induced mutations. As part of the program of introducing new varieties, the Experiment Station in 1987 released 'Arkot 518', a new cotton cultivar especially adapted for irrigation.

The Experiment Station often works with private grants to aid in solving a commercial problem. Tommy Goodwin developed a more efficient method of processing chicken with reduced energy requirements. Allen Canning Company at Siloam Springs worked with John Bowers and M.J. Goode in developing new varieties of spinach. The peach varieties 'Allgold' and 'Goldilocks' developed by James N. Moore are utilized in Gerber's baby food. Jim Brooks and Roy Sharp are working on starch sweeteners. Ron Buescher has developed a process of using calcium to inhibit the softening action of enzymes in fermentation vats for pickles with the effect of reducing salt requirements. Weyerhaeuser Company's Southern Forestry Research Center in Hot Springs works with the Station's forestry research program in joint studies of water run-off patterns and wildlife management.

Soybeans had become the most important cash crop in the state by the 1980s. The Arkansas Soybean Promotion Board, created in 1971 and supported by a 1 cent/bu check-off earmarked for

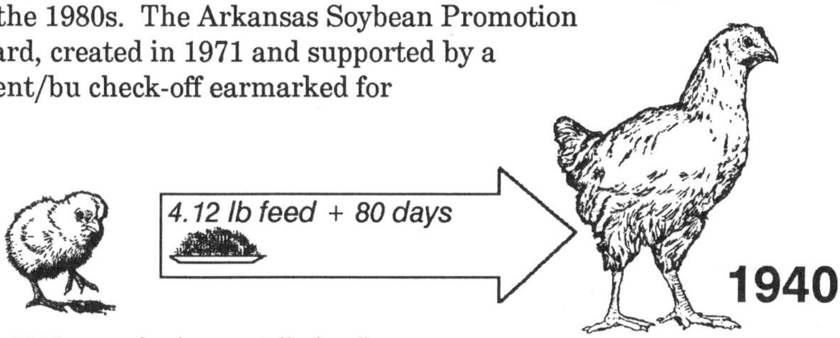

In 1940, producing a 4-lb broiler required 4.12 lb of feed for every pound of broiler and took 80 days. By 1988 only 1.9 lb of feed was required for each pound of broiler, and the process could be completed in 42 days.

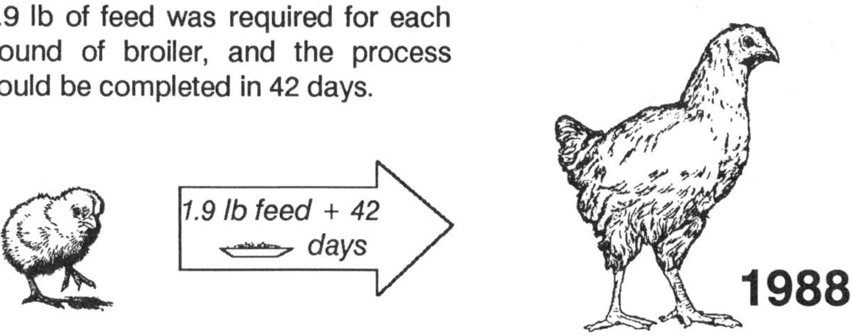

research and market promotion, allowed for even more intense research efforts.

The poultry industry provided for an enormous increase in employment in the state. Companies such as Tyson Foods and Hudson Foods became nationally known producers. The Experiment Station had over 20 scientists working directly or indirectly on problems related to the poultry industry. Two projects of particular interest were Ivan Berry's efforts to develop a low-cost evaporative cooling system for broiler houses and D.R. Davis' efforts to reduce cholesterol in eggs.

The Livestock and Forestry Branch Station under Kenneth F. Harrison is presently involved in several important research areas in cooperation with the Arkansas Forestry Commission and the University of Arkansas at Monticello. The research includes comparing herbicides to ascertain their effectiveness and usage and ascertaining methods of improved forest management for existing tracts as well as the transformation of formerly unproductive land. The station has been conducting experiments with hybrid bermudagrass and fescue as well as with supplements including selenium and zinc and

Raymond W. Barclay Jr. (second from right) was host and producer of *Kaleidoscope*, a monthly program about agricultural news and research, broadcast on the Arkansas Education Television Network from 1977 to 1986. Members of the television production crew are (l to r) Mark Andrews, Earl Robinson, Dale Richardson, Robert Bland, Randy Horn, Barclay and Jimmy Higgins. (Photo courtesy Arkansas Educational Television Network.)

has continued its tests on fertilization of Plains bluestem and Coastal bermudagrasses, treating Coastal bermudagrass and tall fescue hays with anhydrous ammonia and testing varieties of sunflowers in cooperation with the University of Arkansas at Pine Bluff.

The communication and publication of new developments remains a major function of the department of agricultural publications under the auspices of Raymond W. Barclay, Jr. The department has published over 1700 publications, including Bulletins, Report Series, Research Series, Special Reports and the bimonthly journal of *Arkansas Farm Research* as well as producing 700 radio and television programs; the department also has an extensive press release program.

The Experiment Station continues to serve as an important research resource. In 1900, the Experiment Station started an entomological museum, reorganizing it in 1952 for storage purposes. In 1964, the collection, with 37,631 specimens of 4,114 species, was consolidated to provide a quick and ready reference for county extension agents, entomologists and private citizens. C.E. Carlton, the curator, continues to add new species of the state's insect fauna and maintains the collection of approximately 400,000 pinned specimens

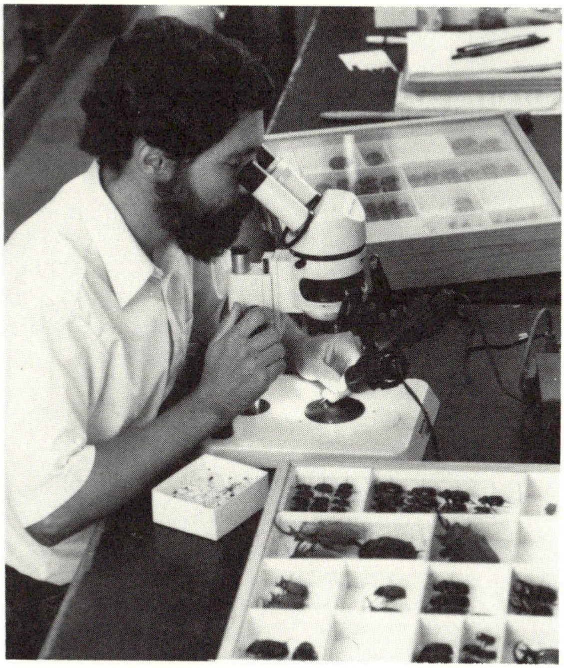

Chris Carlton, curator of the Station's insect collection, inspects one of the more than 500,000 specimens located in the department of entomology. The museum provides faculty, students and residents of Arkansas a resource for insect identification, research and education.

and an additional 100,000 specimens in fluid or on slides representing 30,000 species.

Today, in 1988, as the Experiment Station enters its second century of growth, it faces unprecedented challenges. No longer do the scientists merely serve to test fertilizer and new seed varieties; now they deal with the entire panoply of research that touches agriculture and rural life and the lives of the urban citizenry as well.

The Experiment Stations have evolved to achieve new objectives. The lone scientist has been augmented by teams of specialists. The Experiment Station today concentrates on cooperative research in a regional context as well as within the national scope of the U.S. Department of Agriculture. State-wide and national computer hookups allow for the instantaneous transmission of information. County extension agents have immediate access to the informational matrix developed by the Station on seed varieties, fertilizers, herbicides and planting dates. An informational revolution has given farmers knowledge to operate at optimal levels.

As originally envisaged by the authors of the Hatch Act, Experiment Station scientists would apply their scientific tools to agriculture to deal with immediate needs as well as to disseminate available scientific information to farmers. By the Adams Act, the scientists were also to undertake basic research to pave the way for future breakthroughs. Efforts of pioneer scientists such as C.K. McClelland, Barnett Sure, Dwight Isely and George Becker benefited not only Arkansas, but also farmers and researchers everywhere. The exigencies of the Great Depression brought to bear the analytical tools of the rural sociologists and economists who were funded through the Purnell Act. By World War II, the increasingly complex scientific problems required cooperative research among different scientific disciplines. Establishment of regional laboratories by the federal government assisted the ongoing work of the state-based agency. On the other hand, the growing importance of state branch stations allowed for experimentation and adaptation in a microcosm.

The scientific revolution in agriculture was accelerated by the exigencies of World War II and the technological revolution in the postwar period. Driven by fears of a recurrence of a Great Depression, the Agricultural Marketing Act of 1946 emphasized the need to assist in the processing and sale of commodities as well as production. The economic seesaw of the late 1940s and early 1950s ultimately produced a profit squeeze in the mid-1950s. The Hatch Act of 1955 recommitted the government not only to existing programs but also to crop substitution such as soybeans and beef for cotton and corn. The triple revolution of chemicals, mechanization and technology changed the face of agriculture.

In the 1960s, the growing realization of the possible ecological degradation of chemicals and the mixed effect of the rural exodus caused by mechanization led to the so-called "New Agenda" in scientific research. The Experiment Station found itself having to deal with political as well as scientific questions. Congress widened the competition for grants as well as mandating areas of scientific industry. Research was directed into such non-traditional areas as public health, recreation and retirement. Moreover, the Arkansas Experiment Station adopted programs to monitor water and air pollution, acid rain and pesticides. Agricultural scientists also had to deal with the potential of biotechnology and genetic manipulation. The Experiment Station today combines federal and state research objectives with private concerns to service a broad range of interests.

After 100 years, the Arkansas Experiment Station has contributed to the dramatic economic and social transformation of Arkansas. Research efforts are published and sent throughout the nation and the world. Ongoing efforts help assure the state, nation and world of abundance of food and fiber into the next century.

In spite of the Experiment Station's continued efforts, the state and national trend of farm-to-urban immigration continues. Arkansas' small farms have undergone a major decline in fruit and vege-

To commemorate the Station's Centennial observance in 1988, several projects were planned throughout the state, including centennial gardens of crops, poultry and animal science exhibits and a symposium. The University of Arkansas Museum also prepared a display about the history of the Station. Standing in front of that exhibit are (l to r) former Station directors E. M. Cralley, Lloyd O. Warren and Preston E. La Ferney and current director Gerald J. Musick.

table production, and the number of farmers has declined 80% since the 1920s. Unfortunately, the entire future of the family-based agricultural unit is still very much in doubt in the face of highly efficient corporate agri-business.

Nevertheless, as former director Warren cautioned, farmers must think beyond the temporary surpluses depressing commodity prices to prepare themselves to face the inevitable challenges ahead. As Director Warren viewed it, optimal producer models in soybeans, rice, cotton and other crops are necessary to ensure the best balance among farming inputs. The Arkansas Agricultural Experiment Station has played a major role in this achievement of transforming and reshaping Arkansas in all its aspects and promoting the economic progress of its citizens.

SELECTED BIBLIOGRAPHY

Primary Sources

1. Allan, E. Correspondence of E. Allan. Record Group 164, National Archives, Washington.

2. Arkansas Agricultural Experiment Station. Bulletins, Arkansas Farm Research, Miscellaneous Reports, Cooperative Research Reports, Annual Reports.

3. Records of the Office of Experiment Station Record Group 164. National Archives, Washington.

4. True, Adam. Notes. Record Group 164, National Archives, Washington.

5. United States Department of Agriculture, Office of Experiment Stations. Report on the Agricultural Experiment Station 1890-1954. Washington: U.S. Government Printing Office.

Secondary Sources

1. Alexander, Donald C. 1943. The Arkansas plantation, 1920-1942 New Haven: Yale Press.

2. Baily, Joseph Cannon. 1971. Seaman A. Knapp: Schoolmaster of American agriculture. New York: Columbia University Press.

3. Baker, Gladys L., Wayne D. Rasmussen, Vivian Wiser and Jane M. Porter. 1963. Century of service: The first 100 years of the United States Department of Agriculture. Washington: USDA.

4. Busch, Lawrence and William B. Lacy. 1983. Science, agriculture and the politics of research. Boulder, Colorado: Westview Press.

5. Conover, Milton. 1924. The Office of Experiment Stations: Its history, activities and organization. Baltimore: Johns Hopkins Press.

6. Danbom, David B. 1979. The resisted revolution: Urban America and the industrialization of agriculture 1900-1930. Ames: Iowa State University Press.

7. Daniel, Pete. 1985. Breaking the land: The transformation of cotton, tobacco, and rice cultures since 1880. Urbana: University of Illinois Press.

8. Decanio, Stephen J. 1974. Agriculture in the postbellum South: The economics of production and supply. Cambridge: MIT Press.

9. Dethloff, Henry C. 1970. Rice revolution in the Southwest, 1880-1910. Arkansas Historical Quarterly 29(1):66-75.

10. Dupree, A. Hunter. 1957. Science in the federal government: A history of policies and activities to 1940. Cambridge: Belknap Press.

11. Ford, Arthur M. 1973. Political economics of rural poverty in the South. Cambridge: Bellinger Publishing Co.

12. Hadwiger, Don. 1982. The politics of agricultural research. Lincoln: University of Nebraska Press.

13. Hayes, Ralph W. 1930. A financial study of growing Southern Pines in Hampstead County, Arkansas. U.S. Department of Agriculture Southern Forest Experiment Station, New Orleans, Louisiana.

14. Hightower, Jim. 1972. Hard tomatoes, hard times. Washington Agribusiness Accountability Project.

15. Kerr, Norwood Allen. 1987. The legacy: A centennial history of the state Agricultural Experiment Stations. Columbia: Missouri Agricultural Experiment Station.

16. Knoblauch, H.C., E.M. Law and W.P. Meyer. 1962. State Agricultural Experiment Stations: A history of research policy and procedure. U.S. Department of Agriculture Miscellaneous Publication no. 904.

17. Leflar, Robert. 1972. The first 100 years: Centennial history of the University of Arkansas. New York: Bobbs-Merrill Co.

18. Lavallard, Marie L. 1976. Agricultural progress in Arkansas. Arkansas Agricultural Experiment Station Special Report 28.

19. Marcus, Alan I. 1985. Agricultural science and the quest for legitimacy: Farmers, agricultural colleges and Experiment Stations 1870-1890. Ames, Iowa: Iowa State University Press.

20. McConnell, Grant. 1953. The decline of agrarian democracy. Berkeley: University of California Press.

21. Moore, Ernest G. 1967. The Agricultural Research Service. New York: Praeger.

22. Rosenberg, Charles E. 1976. No other gods: On science and American social thought. Baltimore: Johns Hopkins University Press.

23. Rosenberg, Charles E. 1971. Science, technology and economic growth: The case of the Agricultural Experiment Station scientist, 1875-1914. Agricultural History 45(1):1-20.

24. Rossiter, Margaret W. 1975. The emergence of agricultural science: Justus Liebig and the Americans, 1840-1880. New Haven: Yale University Press.

25. Ruttan, Vernon W. 1982. Agricultural research policy. Minneapolis: University of Minnesota Press.

26. Scott, Roy V. 1971. The reluctant farmer: The rise of agricultural Extension to 1914. Urbana: University of Illinois Press.

27. Shideler, James H. 1957. Farm crisis: 1919-1923. Berkeley: University of California Press.

28. Smith, David C. 1980. The Maine Agricultural Experiment Station: A bountiful alliance of science and husbandry. Orono: University of Maine Press.

29. Street, James H. 1957. The new revolution in the cotton economy: Mechanization and its consequences. Chapel Hill: University of North Carolina Press.

30. Tindall, George B. 1967. The emergence of the new South, 1913-1945. Baton Rouge: L.S.U. Press.

31. True, Alfred Charles. 1937. A history of agricultural experimentation and research in the United States, 1607-1925, including a history of the United States Department of Agriculture. U.S. Department of Agriculture Miscellaneous Publication no. 251.

32. Ware, J.O. n.d. Origin, rise and development of American upland cotton varieties and their status at present. Unpublished manuscript.

Personal Interviews

1. Barclay, Raymond W. Jr., department head, agricultural publications, University of Arkansas

2. Berry, Calvin R., professor, agricultural economics and rural sociology, University of Arkansas

3. Bryan, Billy B., professor, biological and agricultural engineering, University of Arkansas

4. Caviness, Charles, university professor, agronomy, University of Arkansas

5. Goodwin, John, university professor, agricultural economics and rural sociology, University of Arkansas

6. Green, Bernal L., cooperating agent, agricultural economics and rural sociology, University of Arkansas; economist, U.S. Department of Agriculture

7. Lavallard, Marie L., former department head, agricultural publications, University of Arkansas

8. Moore, James N., university professor, horticulture and forestry, University of Arkansas

9. Morelock, Teddy E., associate professor, horticulture and forestry, University of Arkansas

10. Schneider, Mary Jo, adjunct professor, agricultural economics and rural sociology; professor, anthropology, University of Arkansas

11. Templeton, George E., university professor, plant pathology, University of Arkansas

12. Thompson, Lyell, professor, agronomy, University of Arkansas

13. Wheeler, Laurin G., associate professor, horticulture and forestry, University of Arkansas

14. York, J.O., professor, agronomy, University of Arkansas

INDEX

A

B

C

continued

continued

D

continued

E

F

G

J

K

L

M

N

O

P

Q

R

S

continued

T

U

V

W

Y

Z

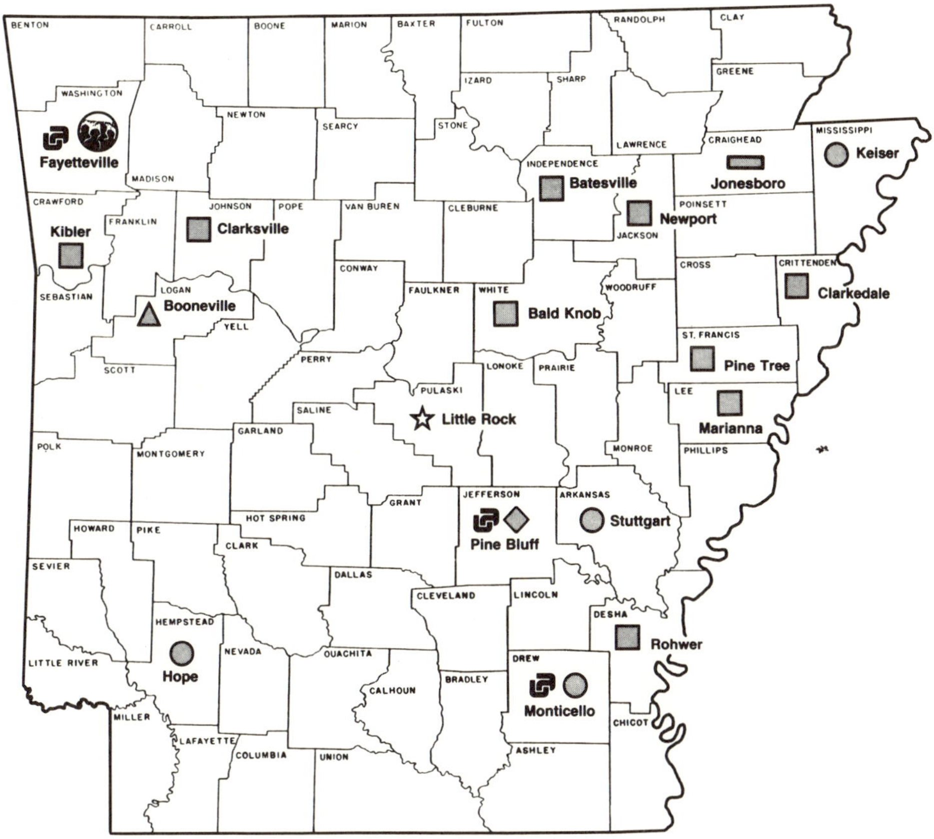

University of Arkansas
Agricultural Experiment Station
Research Units
And Cooperating Agencies

- University of Arkansas Division of Agriculture Instructional & Research Programs
- Agricultural Experiment Station Main Station
- ☆ Cooperative Extension Service State Office
- ◇ UAPB – 1890 Institution
- Arkansas State University Agricultural Sciences
- ○ Research and Extension Centers
- □ AAES Substations
- △ USDA-ARS Small Farms Research Center

Extension Office Located in Each County.

About the Author

Dr. Stephen F. Strausberg

Dr. Stephen F. Strausberg is an associate professor of history at the University of Arkansas, Fayetteville. A native of Brooklyn, New York, he received his B.A. (*cum laude*) in 1964 from Brooklyn College and his Ph.D. in 1970 from Cornell University. He joined the faculty of the University of Arkansas in 1968 after a year as a researcher with the U.S. Public Land Law Review Commission. His research fields include American agriculture, U.S. economic history, public domain and Arkansas in the Depression. He resides in Fayetteville with his wife and two daughters.

Notes

Notes

Notes